OUR LIVING WORLD OF NATURE

The Life of the Far North

Developed jointly with The World Book Encyclopedia

Produced with the cooperation of
The United States Department of the Interior

OUR LIVING WORLD OF NATURE

The
Life
of the
Far North

WILLIAM A. FULLER
and
JOHN C. HOLMES

Published in cooperation with
The World Book Encyclopedia

McGraw-Hill Book Company

NEW YORK   TORONTO   LONDON

WILLIAM A. FULLER *is a graduate of the University of Saskatchewan and earned his Ph.D. at the University of Wisconsin. After several years as a research biologist with the Canadian Wildlife Service, he joined the faculty of the University of Alberta and is now Chairman of the Department of Zoology. He has served on the boards of directors of the International Union for Conservation of Nature, the Canadian Audubon Society, the National and Provincial Parks Association of Canada, and the Canadian Society of Wildlife and Fishery Biologists.*

JOHN C. HOLMES *did his undergraduate work at the University of Minnesota and received the M.S. and Ph.D. degrees at Rice University. A specialist in the ecology of parasitic helminths, he is a professor of zoology at the University of Alberta, where he teaches courses in boreal and marine ecology. He is a member of the Arctic Institute of North America, the Canadian Society of Zoologists, the American Society of Mammalogists, and other scientific organizations. Like author Fuller, he returns to the far north as often as possible.*

Library of Congress Cataloging in Publication Data

Fuller, William Albert, 1924–
    The life of the far north.

    (Our living world of nature)
    1. Tundra ecology—Arctic regions. 2. Taiga ecology—Arctic regions. 3. Natural history—Arctic regions. I. Holmes, John C., 1932–   joint author.
II. Title.
QH541.5.T8F84   574.5′264   78-38783
ISBN 0-07-022614-8
ISBN 0-07-046014-0 (sub. ed.)

1 2 3 4 5 6 7 8 9 10 NR RM 7 8 7 6 5 4 3 2

Contents

LAND OF THE LONG DAY   *155*

APPENDIX

The Northern World

Canoe travel is easy on the Peace River, a broad expanse of sparkling water that winds its way northeastward across the province of Alberta in western Canada. In the main channel the current swirls in giant eddies that twist the canoe about, but away from the channel, the current is weaker, and the water is as smooth as a sheet of glass. Here you can drift lazily with the flowing water, using the paddle only to maintain your course. Without working very hard, you can travel two or three miles an hour and still have plenty of opportunity to observe the scenery and wildlife on the river and along its banks. It is an ideal place to begin a journey through the great north woods toward the North Pole.

The scene along the river is one of wild solitude with scarcely a sign of human life. The farms and grainfields of central Alberta are far behind you now. Here you are surrounded by an immense, gently rolling landscape of vast forests dotted with grassy marshes, bogs, and glistening lakes and ponds. In some places, willows and alders replace the taller trees. They are especially common on mudflats on the downstream ends of islands, in spongy soil on the inner banks of curves in the river, and in other damp places. On higher, drier ground the river is bordered by large patches of quaking aspens, their small leaves fluttering silver in the breeze. Still other areas are covered with mixed stands of aspens and spruces. But most of the forest is a great unbroken expanse of white spruces, their neat conical crowns rising to heights of eighty or one hundred feet.

When we visualize the north woods, we tend to think mainly in terms of *coniferous*, or cone-bearing, trees such as spruces and other evergreens. However, all these different kinds of forest beside the river are parts of the general vegetation type known as northern coniferous or *boreal* (northern) forest. An even better name, emphasizing the unique character of these mixed forests typical of northern regions, is a Russian word, *taiga*, pronounced "*tie*-ga."

In North America a broad band of taiga stretches across much of Canada and sweeps west through most of central Alaska. In the eastern half of the continent taiga merges on the south with mixed forests of conifers and broadleaf trees such as maples, birches, and beeches. However, large islands

At Virginia Falls in Canada's Northwest Territories, the sparkling water of the South Nahanni River accentuates the somber tones of the vast northern evergreen woodland known as the boreal forest, or taiga. A powerful erosive force, the river here plunges down a cataract twice the height of Niagara Falls.

of typical boreal forest also exist farther south in northern Maine; on the mountains of New Hampshire, Vermont, and New York; and in scattered patches around the Great Lakes. Across the Canadian plains, in Manitoba, Saskatchewan, and Alberta, the taiga is bordered on the south by aspen parklands which give way in turn to prairies, while in western Alberta and British Columbia, the boreal forest merges in the mountains with coniferous forests made up of different species of evergreens such as Douglas fir and Engelmann spruce.

Nor is taiga restricted to North America. Very similar forests stretch in a nearly continuous belt all the way around the northern part of the world. The trees in the taiga of Scandinavia and the Soviet Union are different species from those found in North America, but the appearance of the trees and of the forest as a whole is remarkably similar everywhere around the northern hemisphere. The distribution of taiga therefore is described as *circumpolar*, which literally means "around the pole." As you continue on your northward journey, you will encounter many examples of circumpolar plants and animals.

Although the taiga is made up mostly of unbroken expanses of spruces and other conifers, boreal forests also include extensive tracts of deciduous trees, primarily aspens. Like deciduous forests farther to the south, the mixed boreal forests provide a spectacle of color in autumn.

Life along the river

Drifting north with the current, you probably will notice relatively little wildlife at first. It is June and the midday sun has sent the thermometer to over 90 degrees—most of the larger animals probably are resting in the shade. Overhead, however, a pair of red-tailed hawks circles lazily on broad, rather blunt wings. As they bank and soar, the sun occasionally highlights their distinctive rusty red tails. Along the riverbanks, delicate sparrow-sized spotted sandpipers teeter on long wire-thin legs. Ducks also are common, especially mallards, widgeons, green-winged teal, lesser scaup, and common goldeneyes. They seem especially fond of lingering among the tangled trunks and branches of fallen trees washed up against the shore.

Tracks on muddy or sandy stretches of riverbank hint of the activities of many other creatures as well. One large track, which looks surprisingly like a human footprint, is the mark of a black bear, an adaptable beast that thrives here as well as in more temperate regions. From time to time as you round a bend, you may surprise a cow moose and her

Seeking refuge from danger, a young black bear scrambles up the stout trunk of an aspen. These wide-ranging, adaptable bears live throughout the North American taiga, but also are found in wild areas as far south as Florida and central Mexico.

gangling calf drinking at the water's edge, or perhaps a bull moose munching on the greenery. Red foxes also patrol the river banks. Occasionally they den beneath overhanging roots of trees where the river in flood sometimes excavates caverns.

If you are really lucky, you may see another set of dog-like tracks, much larger than those of foxes, for this country is inhabited by wolves. When you see the wolf tracks, you will know for certain that you are in the wilderness, since wolves rarely survive near inhabited areas, although they sometimes live quite close to towns and cities in Alaska. At one time they ranged through much of the United States, even as far south as Florida. But as settlers and ranchers advanced across the continent, they killed off the wolves and drove them farther and farther into the wilderness. Now all the wolves are gone from the United States, except in Alaska, in Isle Royale National Park in Lake Superior, and in a few scattered areas in northern Minnesota, Wisconsin, and perhaps in a few other places. Nor have they fared much better elsewhere, for man's hand has been turned against wolves for centuries. They have been completely

To catch a glimpse of the gray, or timber, wolf is a rare treat for any visitor to the far north. The magnificent animals, which may weigh one hundred pounds or more, are scarce, however, and usually stay far from human beings.

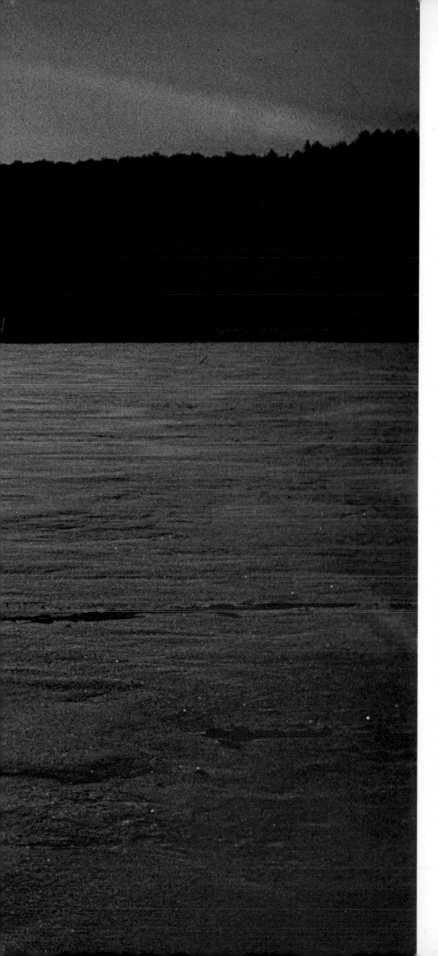

THE WOLVES OF ALGONQUIN PROVINCIAL PARK

In winter's long twilight, three wolves gambol across the snowy surface of a frozen lake in Ontario's Algonquin Provincial Park. The park is a vast wilderness tract of lakes and forests in southeastern Ontario, within easy reach of both Toronto and Ottawa. Many animals of the taiga find refuge in the park, but wolves are a prime attraction. Although few visitors ever actually see the wolves, many are rewarded by the sound of their howling in the night. The eerie beauty of their melodious chorus floating across the evening landscape is enough to convince most people that wilderness must be preserved.

eradicated from Britain and much of the rest of Europe, are scarce in Scandinavia, and are persecuted throughout the Soviet Union. Even in the Canadian wilds they have been faced with poisoning campaigns based on the unproved assumption that fewer wolves would mean more moose and caribou.

In recent years, fortunately, some people have come to the wolf's defense. A wilderness without wolves, they say, is no wilderness at all. They say that the time has come to stop persecuting the world's few remaining wolves, to let them survive unmolested in the last few tracts of northern wilderness. Anyone who has ever glimpsed a wolf in the forest is likely to agree. Perhaps even more dramatic is the sound of their eerie howling chorus in the evening. To hear their wild music is an experience never to be forgotten. It has the power to convince almost anyone that the world is big enough to harbor wolves as well as people.

The spruce forest

To find out what the taiga is really like you must leave the river and go into the forest itself. Once you step beneath the towering canopy of spruces, you will find that living conditions in the forest are far different from those outside. For one thing, it is much cooler than it was on the river, and instead of blazing sunlight you are now in dim shade. If you look up, you will see that the needle-covered branches of adjacent trees intermingle so thoroughly that they form an almost unbroken canopy.

As a result, very little sunlight gets through to warm and light the forest floor. In addition the air is very calm. Although the tips of the trees nod with passing breezes, the dense foliage practically eliminates any air movement at ground level. Finally, the forest floor seems parched and dry. Again, the dense canopy of branches is responsible. Even in a heavy rainfall, most of the rain is caught and held by the needles, where much of the water later evaporates without ever reaching the forest floor.

Obviously such cool, dim, dry living conditions are not very favorable for most plants. If you are familiar with mixed hardwood forests like those in the eastern United States, with their profuse undergrowth of shrubs and flowers,

Wintergreen is one of the few plants able to survive in the cool, dim environment on the floor of the spruce forests of the taiga. Its bright red berries provide a welcome source of winter food for many animals.

the spruce forest is likely to strike you as rather barren. Even the lower branches of the trees are dead brittle skeletons: as the spruces grow taller, the lower branches are gradually shaded out by new growth overhead and die.

Beneath each spruce is a thick mat of dead needles, while the rest of the forest floor is covered with a nearly continuous bed of feather mosses. A few small plants have managed to get a foothold in the moss: twinflowers, wintergreen, bunchberry, and horsetails. The only shrubs are widely scattered alders, wild roses, and buffalo berries, while here and there a young spruce is struggling up to reach the light.

Sometimes called the "fool hen" because it shows so little fear of man, the spruce grouse can easily be approached for a closer look. Although its main foods are the needles and buds of conifers, it sometimes varies its dict with berries, mushrooms, seeds, and even insects.

Animals of the spruce forest

You are certain to notice, too, that the spruce forest is a relatively quiet place. It is not enlivened by many bird songs, since relatively few animals live in such an unvaried *habitat*, or living place. Yet some animals specialize in the foods provided by spruces and other conifers. The most unusual is the spruce grouse, one of the few animals that eat spruce needles. These handsome birds are often difficult to locate,

With the help of their agile forepaws, red squirrels nimbly rip apart the cones of spruces and other conifers to get at the nutritious winged seeds hidden between the scales. In years of good cone production, these notorious hoarders sometimes store enough cones to carry them through two winters.

for their variegated plumage provides perfect camouflage on the forest floor and among the lower branches of the trees.

A number of birds also eat the seeds of spruces and other evergreens. Streaky, sparrowlike pine siskins depend on the seeds of conifers and other plants in winter, although in summer they eat buds and insects as well. Pine grosbeaks also eat conifer seeds. The most interesting of these birds are red crossbills, with sleek brick-red plumage, and rose-colored white-winged crossbills. At first glance the crossbills seem deformed, since their beaks cross at the tips. In fact, their awkward-looking bills are uniquely adapted for extracting seeds from cones. The birds insert their bills between the scales of a spruce or pine cone, and when they close them, the tips cross and pry the cone scales apart. The birds are then able to use their tongues to extract the seeds hidden at the bases of the scales.

Although not many berries are available for fruit-eating

birds, there are plenty of insects in the spruce forest. In addition to hordes of mosquitoes and flies, which seem to thrive in summer everywhere in the far north, the trees also harbor creatures such as the larvae of spruce bud worms and larch sawflies. Red-breasted nuthatches, boreal chickadees, golden-crowned and ruby-crowned kinglets, myrtle warblers, and many other insect-eating birds are on hand to enjoy the insect feast. On dead or injured trees, three-toed woodpeckers drill into the trunks to get at plump insect larvae that bore beneath the bark.

A few mammals also are present. As you walk through the woods, a snowshoe hare may bound away at your approach. The hares leave well-worn trails in the moss of the spruce forest, although they usually feed in brushy areas such as in aspen and mixed forests. Moose also seek the spruce forest for shelter in winter, even though they find little to eat here.

The most characteristic mammals, however, are red squirrels. Wherever you go in the coniferous forest, you will be greeted by their nervous chattering calls. Even if you do not see them, you are sure to notice signs of their feeding activities. Heaped around their favorite feeding stations are piles of the remains of spruce cones. The squirrels nip off the cone scales, one by one, to get at the seeds, until nothing is left but the twiglike central cores of the cones.

Red squirrels in turn are preyed upon by martens, sleek, slender members of the weasel family. Despite their size— a male may be as much as thirty inches long—martens are lithe and agile as they pursue squirrels through the springy canopy of spruce boughs. Like mink, fishers, and other members of the weasel family, they are valued for their lustrous fur. In fact, they have been so heavily trapped that they are now extremely scarce in some areas.

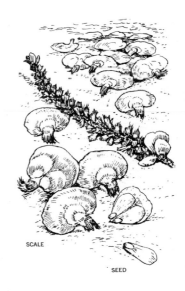

SCALE

SEED

Red squirrels feed on spruce cones by nipping off the scales one at a time to get at the two winged seeds growing at the base of each scale. When the squirrels finish, only the bare central cores of the cones remain. At favorite feeding sites, discarded scales and cores accumulate in heaps known as middens.

Under the aspens

The areas of aspen woodland provide a striking contrast to the spruce forest. The tall naked tree trunks are topped by crowns of fluttering leaves. In dense stands, the branches intermingle to form a continuous canopy, but it is not like the solid mass of foliage in the spruce forest. Much more sunlight filters through or is reflected to the forest floor, and rain also penetrates the open aspen canopy more easily.

The result is a lighter, warmer, damper habitat, where a great profusion of shrubs is able to grow. Mooseberries, saskatoons, and red osier dogwoods all are common, while lower shrubs such as wild roses and snowberries also flourish. On the forest floor pea vines, wintergreens, twinflowers, grasses, and many other plants form an almost continuous carpet. There are no mosses, however, probably because it is too sunny here for them. In any case, the sudden fall of leaves from the trees and shrubs each autumn would cover them like a blanket and literally smother them.

The annual leaf fall in the aspen forest has a lot to do with the wealth of shrubs growing there. In the spruce forest much of the ground was covered with a thick mat of needles. The mat was thick because spruce needles decompose slowly. Relatively few fungi and other decomposers are able to break down the needles of conifers. Moreover, they work slowly in the cool habitat of the spruce forest floor. Thus, minerals are locked up for a long time in the mat of dead needles, and the soil, as a result, is low in nutrients. In addition, when the minerals are released into the soil, they are soon washed down to deeper layers, further impoverishing the soil.

Unlike spruce needles, the fallen leaves of aspens are eaten by a great variety of animals. They are quickly attacked by a variety of insects and other soil animals before fungi and bacteria take over to produce complete decomposition. The result is a thick layer of nutrient-rich humus that supports a much more luxurious and varied growth of plants.

If you search carefully among this dense undergrowth, however, you will find no aspen seedlings. The reason for this is the fact that aspen seeds germinate and the young trees grow only in bright sunlight. Even the light shade cast by mature aspens is too much to permit their growth. Their need for bright sunlight, in fact, explains the existence of large tracts of aspens in the taiga, for aspens are among the first trees to move into areas that have been burned over by forest fires. And forest fires are quite common in the far north. The dry resinous spruce needles are highly flammable and easily ignited by lightning. In addition, many fires are set by human beings, through both carelessness and misguided attempts to clear the land.

Forest fires caused by both nature's lightning and man's carelessness are a constant threat to the taiga. Occasional fires may be beneficial, however, since they release nutrients stored up in the mature spruces and create openings where new forests can develop.

New forests from old

When the fires die out, the scorched earth is colonized at first by fireweeds and other rugged *annuals* (plants that grow from seedlings to mature seed-bearing plants during a single growing season, then die). Aspen seeds eventually blow in on the wind and gain a foothold in these brightly lighted, weed-filled tracts. Gradually they grow up to produce a mature aspen forest.

Even though the aspens are not replacing themselves, however, the forest is not doomed to disappear. Here and there beneath the trees you will find white spruce seedlings growing up through the shrubs. In contrast to aspens, the wind-borne spruce seeds germinate poorly in bright light and benefit from the shade provided by the aspens. Over the course of time, if the area is not again ravaged by fire, the spruces gradually overtop the aspens and shade them out. Instead of the once-a-year fall of aspen leaves, spruce needles now rain down constantly and accumulate on the forest floor. In the cooler, darker conditions, mosses are able to grow and form a carpet that inhibits the rooting of many

Fireweed is one of the first and most spectacular annual weeds to take over on land laid bare by forest fires. Sun-loving aspens soon will follow, creating a new, but different kind of forest where an unbroken stand of spruces once covered the land.

herbs and shrubs. Finally the area is once again covered by typical mature spruce taiga, thus completing the process of *succession,* as this gradual replacement of one kind of plant community by another is termed.

During the process of succession from aspen to spruce, there will be a period of mixed forests. Although mixed forests combine some conditions of both spruce and aspen forests, they also have some distinctive features. For example, when both leaves and needles are decaying together, decomposer organisms produce certain organic compounds that are not present in either pure rotting needles or pure rotting leaves. Some of these compounds provide an energy source for new kinds of decomposers that in turn produce a distinctive kind of humus found only in mixed forests. The new kind of humus favors new kinds of plants that did not grow or were rare in either type of pure stand.

Animals also are favored by mixed stands. Food is abundant and varied in the shrubby growth beneath the aspens, while the patches of spruce provide good shelter for nesting or denning and for escape from predators. Ruffed grouse, for instance, are more common in mixed forests than in

Eventually spruces take root in the shady aspen forest. Just as aspens crowded out the fireweed and other annuals, the spruces will gradually overtop the aspens and complete the process of succession from bare ground to mature spruce forest.

either pure aspen or pure spruce forests. In the far north, black bears also seem to prefer mixed forests.

For animals such as moose and snowshoe hares, the mixed forests offer the best of two worlds, with young aspens and shrubs providing plentiful food, and scattered spruce stands providing shelter, especially from the winter cold. All in all, the mixed taiga is a much more cheerful and lively habitat than either the somber pure spruce taiga or the airy aspen forest.

Wilderness sanctuary

The lower Peace River passes for nearly two hundred miles through Wood Buffalo National Park, located near the northeastern corner of Alberta and extending across the border into the Northwest Territories. This huge park of 17,300 square miles, established in 1922, encompasses vast areas of taiga and open plains. It was set up mainly to provide a refuge for wood buffaloes, or wood bison, as they are more properly termed. Wood bison are simply a race of the familiar bison of the western plains. They differ from plains bison only in being slightly larger and somewhat darker, and in their preference for a wooded habitat.

Like the plains bison, they were heavily hunted in the nineteenth century. By 1893, only about five hundred of them remained, in the area around the junction of the Peace and Slave Rivers. Following legal protection and establishment of the park, however, the population gradually recovered. Unfortunately, misguided officials introduced several thousand plains bison into the park. They interbred with the wood bison, so that probably there are no pure wood bison left. But the offspring are hardy, and although it is difficult to estimate the number of animals in such a large area, it is believed that the park may now contain about fifteen thousand bison. To see a herd of them scrambling down the muddy banks of the Peace River, swimming across, and then disappearing on the far shore is an impressive sight—

In Canada's Wood Buffalo National Park, a thousand-pound wood bison hovers protectively over her day-old calf. Once nearly exterminated by hunters, the bison herd now finds complete protection in the 17,300-square-mile park.

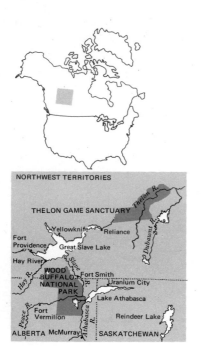

Two huge wilderness areas in western Canada have been set aside to protect endangered animals. Wood Buffalo National Park was created as a sanctuary for wood bison, while the nearby Thelon Game Sanctuary was established as a refuge for muskoxen.

24

and a forceful reminder of the value of preserving wilderness areas.

This is especially true in an area of varied habitats like Wood Buffalo National Park, since far more than bison have flourished in the reserve. Beavers and muskrats are plentiful in the ponds that dot the park. Bears roam through the forests, foraging for any kind of food they can find, whether plant or animal. Moose thrive here; woodland caribou are found in the northern part of the park; and white-tailed deer reach their northern limit in the park. All sorts of ducks also fly north in tremendous numbers to nest in the park.

Strangely enough, this park, which was established to preserve the nearly extinct wood bison, has also proved to be a sanctuary for one of the rarest birds in the world. Although whooping cranes were never really numerous, now

less than one hundred of these stately white birds remain alive in the wild. To preserve this small remnant flock, Aransas National Wildlife Refuge was established on the Gulf Coast of Texas. Here the birds congregate each autumn to winter in closely guarded solitude, and leave each spring to breed in the far north. Every autumn conservationists await the flocks' return to find out how the birds have fared on their breeding grounds.

For years, however, no one knew exactly where they went to breed. Finally, in 1955, author William Fuller found a pair of whooping cranes nesting in Wood Buffalo National Park. Apparently the birds had been nesting there for years. But this pristine wilderness is so vast that no one had ever before discovered that this was where the birds returned each spring to complete their chore of rearing families.

Among the more spectacular residents of Wood Buffalo National Park are moose. They feed mainly on the leaves and twigs of willows and aspens but, like this cow and calf, also wade in marshes and shallow lakes to forage for aquatic plants. The calves, which weigh twenty to twenty-five pounds at birth, are able to walk within a few hours. They remain with the mother for a year or more, roaming the uplands in summer and returning to more sheltered lowland areas in winter.

A RARE BIRD'S REMOTE NESTING PLACE

Great areas of Wood Buffalo National Park are covered with marshes and shallow ponds fringed with stands of black spruce. It is here that the world's few remaining whooping cranes return to nest each summer.

Although never really abundant, whooping cranes once nested over a wide range extending from the Northwest Territories to Iowa and northern Illinois. As the majestic birds, now one of the rarest species in the world, slipped to the brink of extinction early in this century, the breeding locality of the few survivors became a major biological mystery. Every autumn, the meager flock returned to its wintering area on the Texas coast, where the Aransas National Wildlife Refuge was established in 1937 for the birds' protection. But no one knew where the cranes went in summer until 1955, when author Fuller and a companion spotted one of the birds sitting on its nest in Wood Buffalo National Park as they flew overhead in a small plane. Since then, many more nesting pairs of whooping cranes have been observed in this unique wilderness sanctuary.

Pure white except for their black wingtips and dark red faces and crowns, a pair of whooping cranes (right) stands out in sharp contrast to the brushy surroundings in Wood Buffalo National Park. Below, a family of the elegant birds traces a lacy pattern of footprints on the muddy bottom of a pond in the park.

Boreal woodlands

The next stop on our journey to the far north is at Great Slave Lake in the Northwest Territories just a short distance to the north of Wood Buffalo National Park. It is an area of spectacular scenery. The eastern end of the lake is surrounded by rolling hills of pinkish and grayish granite. The lake is dotted with rocky islands of literally every size and shape. Some of them barely rise above the water, while others tower up as high as six hundred feet in sheer cliffs. Some of the peninsulas jutting into the lake are bordered by even taller rocky cliffs.

The islands and rocky hills to the east of the lake are part of a geological formation called the Precambrian Shield that is widespread over the north-central part of North America. Precambrian means that it was formed before the Cambrian period began about 600 million years ago. Indeed, rock samples taken from the north shore of Great Slave Lake are known to be more than two *billion* years old, which makes these ancient granites some of the oldest rocks found anywhere on earth. Geologists tell us that all the continents have cores or shields of similar ancient rocks.

The forest also is quite different from anything you have seen before. As you traveled north from the Peace River area, you might have noticed the forest gradually, almost imperceptibly, changing in character. It would be difficult to pinpoint exactly where the new kind of forest began, for there are few abrupt, clearly defined boundaries in nature. Yet by the time you arrive at Great Slave Lake, you will find that you have left the dense unbroken spruce forest far behind.

The trees are the same species you saw farther south: white spruce, black spruce, and, to a lesser extent, jack pine. Scattered among the conifers are a few aspens and paper birch. But the overall effect is one of an open park land. Instead of forming a continuous canopy, the trees are rather widely spaced, with open areas in-between. In some places shrubs grow beneath the trees, but elsewhere there is little

In some places the dense cover of the boreal forest is broken by openings such as this bog with spruces and larches growing from a carpet of reindeer moss. Farther north the forest thins out to an even more parklike appearance.

31

more than widely separated trees underlain by dense carpets of curious grayish-green plants commonly known as reindeer moss. Since the treetops do not form a continuous canopy, many botanists prefer not to call this formation a forest. Instead they use the term *boreal woodland*. However, most scientists agree that these woodlands are a subzone of the taiga.

Like the rest of the taiga, boreal woodlands are circumpolar in distribution. All the way across North America, a belt of boreal woodland lies to the north of typical taiga. In Eurasia too, the dense taiga gradually grades into boreal woodland. However, in eastern Siberia the trees in the woodland are larches while, for some unknown reason, the comparable forest in northern Scandinavia is dominated by birches.

Lots of lichens

Reindeer moss is so important in the far north that it deserves a closer look. These short, upright plants carpet the ground, their branching arms interlacing to form a mat that resembles stiff, tangled grayish wool. Actually they are not

Hugging the ground, a spruce grouse chick rests on a bed of reindeer moss. The "moss," really an upright lichen, also is known as caribou moss, or caribou lichen. As with all lichens, each plant is a symbiotic pairing of two partners, a fungus and a chlorophyll-bearing alga.

mosses at all but *lichens*, even though they are commonly called reindeer moss or caribou moss.

In all there are about fifteen thousand different kinds of lichens in the world. Each one is unique in being not one plant, but two plants living in close association. The main mass of a lichen is a fungus, a kind of plant that lacks chlorophyll and so cannot manufacture its own food. Bound up in the meshes of fungal threads are many single-celled green or blue-green algae, which do contain chlorophyll and thus can make organic compounds by photosynthesis.

Although in a few cases both the fungi and the algal cells can live independently, each partner in the lichen seems to benefit from their living together. The fungi absorb and store large amounts of water, which the algae need for photosynthesis. The fungi in turn take some of the food that the algae manufacture. Clearly this is a case of *symbiosis*, a situation in which two different organisms live in close association, and probably is the kind of symbiosis called *mutualism*, with both partners benefiting from the relationship. It is a highly successful arrangement, for lichens occur in a wide variety of habitats, thriving in deserts, on bare rocks and mountaintops, and here in the far north.

If you look around, you will discover that lichens grow in several different forms. Upright, branching, shrubby varieties like the caribou lichens are described as *fruticose*. The colorful blotches of red, orange, yellow, and other hues on the otherwise bare rocks are *crustose* lichens, which grow as thin crusts adhering firmly to the rocks. In some places, loose leaflike *foliose* lichens also dot the rock surfaces. Finally, long stringy forms hang like greenish beards from the branches of many of the trees, while crustose and foliose lichens grow on their trunks.

Lichens play several important roles. Caribou lichens and, to some extent, the forms that hang from trees are the main winter foods of caribou in North America and reindeer in Eurasia. Winter survival of caribou, in fact, depends so closely on the abundance of lichens that biologists have made careful estimates of the amounts available. They have found that in typical boreal woodlands there may be four hundred to five hundred pounds of lichens on an acre of forest floor. Lichens grow so slowly, however, that new growth each year often amounts to only three or four pounds per acre. Thus, unless the caribou wander about

Lichens of many kinds grow in the far north. Here a rock is decorated with colorful, crustlike patches of crustose lichens and curling, leaflike foliose lichens.

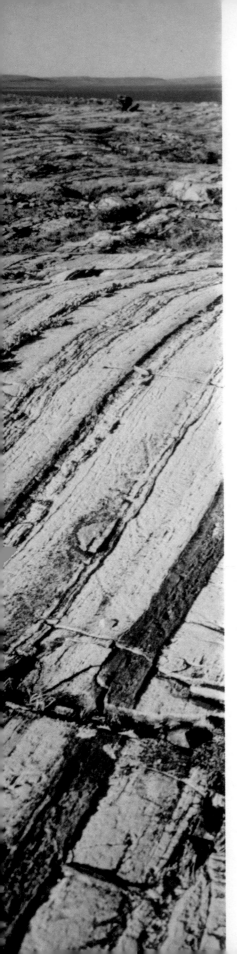

and feed lightly on the lichens over large areas, they can gradually destroy their own food supply.

Even when caribou do not overgraze on lichens, their food supply is often faced with total destruction. When dry, lichens burn easily. The resins in the spruces and accumulations of their dead needles add to the severity of fires in boreal woodlands. Once a fire sweeps through such an area, the area remains useless for caribou for a long time. Because of the lichens' slow growth rate, recovery of a lush lichen mat after a fire frequently requires between one and two hundred years.

Shallow soil

An equally essential role of lichens is the part they play in forming soil. Crustose and foliose lichens produce acids that slowly dissolve parts of the rocks on which they grow. Even more important, they cover the rocks with a layer of plant life, which tends to keep the rock surfaces moist. Chemical weathering is thus made possible, and this is the chief way in which granite rocks get broken down. In addition, tiny bits of wind-blown sand and plant fragments are trapped on the lichen-covered rocks. In time a primitive sort of soil is formed that can support mosses. As the mosses grow and the layer of vegetation thickens, weathering takes place at a faster rate. Eventually there is enough soil for more advanced plants such as shrubs and even trees to get a foothold, thus completing the succession from bare rock to forest.

Yet as you look around, you may well wonder why, if these Precambrian rocks are so old, they are not covered with a much thicker layer of soil. In the distant past they probably were. But in the last million years or so, the Canadian Precambrian Shield was covered four times by continental glaciers. Even today, most of Greenland is covered by a great ice cap, and glaciers still creep down from the mountains in Alaska, western Canada, and elsewhere.

They are grim reminders of the potent force of the vast,

Thousands of years ago a passing glacier scraped these Canadian rocks bare of soil and polished their surfaces into smoothly undulating waves. Debris has begun to collect in depressions, however, and eventually the rocks may once again be covered with soil.

34

GLACIER BAY NATIONAL MONUMENT

At Glacier Bay National Monument, a 4,400-square-mile preserve on the southeastern coast of Alaska, visitors can witness at firsthand the awesome spectacle of glacial ice. Fed by snow falling on the mountains of the Fairweather Range, sixteen active glaciers, such as Margerie Glacier (*left*), sweep down the steep slopes and plunge into the sea. A popular tourist excursion in the park is the tour aboard a National Park Service patrol boat, which permits close-up, water-level views of the glaciers (*right*). On the next two pages, seagulls hitch a ride on an iceberg that broke loose as the snout of a glacier slipped into the sea.

A multiple-exposure photograph traces the path of the midnight sun across the arctic sky over Bylot Island, well within the Arctic Circle in the Northwest Territories. The picture was shot at fifteen-minute intervals . . .

mile-thick sheets of moving ice that once ground across most of Canada and into the northern United States. If you look at some of the exposed granites around Great Slave Lake, you will see that many of them are scarred with northeast-southwest lines. These scratches were gouged into the rocks by boulders that were carried beneath the tremendous weight of glacial ice. As the glaciers advanced, they easily swept away any soil that covered the rocks. All the thin mantle of soil now covering the rocks has been formed in the past six thousand years, since the final retreat of the last glaciers from the area around Great Slave Lake.

Land of the midnight sun

While this poor soil development has much to do with the relative sparsity of trees in boreal woodlands as compared with typical taigas, other factors are operating as well. As you travel north in early summer, it is impossible not to notice that the days grow progressively longer. As you canoed along the Peace River, the sun did not set until ten o'clock in the evening, and the first light of dawn began to show at two or three o'clock in the morning. Around Great Slave Lake, just a few hundred miles farther north, the mid-summer nights are even shorter, for the far north, as everyone knows, is often called the land of the midnight sun.

This odd phenomenon results from the fact that the earth

is tilted at a 23½-degree angle on its axis. As a result, when the earth proceeds on its annual orbit around the sun, the North Pole is angled toward the sun during the northern summer and away from the sun in winter. Because of this, the sun never sets on the North Pole in summer, and never rises there in winter.

The southern limit of twenty-four-hour daylight in summer and twenty-four-hour darkness in winter is the *Arctic Circle*. This is the imaginary line around the earth at 66½ degrees north latitude, or 23½ degrees south of the North Pole, which exactly corresponds to the angle of tilt of the

. . . from 10:30 P.M. to 1 A.M. on July 25. At midnight (the seventh exposure) the sun dipped closest to the horizon, then began to rise again over a landscape where in summer the sun never sets and in winter never rises.

earth's axis. At Great Slave Lake, the Arctic Circle is still a few hundred miles to the north. But in June, as you approach this imaginary circle, the period each night when the sun is below the horizon becomes shorter and shorter. On the Arctic Circle, the sun never sets on the first day of summer, the longest day of the year (about June twenty-first). But the midnight sun lasts only one night. As you go north of the circle, the summer "day" of twenty-four hours of sunlight lasts longer and longer, until at the Pole it continues uninterrupted for the six months between the spring and fall equinoxes.

In winter, the opposite thing happens. On the first day of winter (about December twenty-first), the shortest day of the year, the sun never rises above the horizon of the Arctic

41

Circle. At the circle, the winter "night" lasts just one day, but as you go north, the "night" is longer and longer, until at the North Pole it lasts for six months.

The same thing happens at the South Pole, but at opposite times of year. During the northern summer, the South Pole is immersed in its long winter night, while it enjoys its season of twenty-four-hour sunlight during the northern winter.

The result of all this is an increasingly severe climate that drastically limits plant growth. Over the course of the year relatively little sunlight is available to warm the far northern landscape. Long, dark, bitterly cold winters alternate with brief, cool summers. At the eastern tip of Great Slave Lake, the highest temperature ever recorded was 90 degrees, but that was exceptional. In July, the warmest month of the year, the average temperature there is only 55 degrees. And in January the average temperature at the east end of Great Slave Lake is 21 degrees below zero!

In addition to the short growing season and low temperatures, plant growth is further inhibited by lack of moisture. Combined rainfall and snowfall over much of the far north is so low that the whole area can be considered a semi-desert. Counting ten inches of snow as equaling one inch of rain, the total annual precipitation here amounts to a mere ten inches. This is about the same as the annual rain-

Beyond tree line, the forests of the taiga give way to the vast treeless expanses of the tundra. Here, on the Yukon Delta in Alaska, grassy plains are pockmarked with shallow pools, but elsewhere large areas of the tundra are covered with coarse gravelly soil that is nearly devoid of vegetation.

fall in many of the deserts of the American Southwest. In New York City, in contrast, the total rainfall each year averages about forty-two inches, while in tropical rain forests it may exceed one hundred inches.

From tree line to tundra

Just a little to the north of Great Slave Lake, the increasingly rigorous living conditions take their final toll on the growth of trees. At its northern extreme, the circumpolar belt of taiga is bordered by *tree line,* the point beyond which trees are unable to grow.

You should not expect to step from the forest immediately into a treeless area, for the boundary is not a rigid one. Instead, the trees gradually thin out until they appear only in widely scattered groves in especially well protected places. Some of the groves may cover an acre or two, but others comprise only a dozen or so black or white spruces.

The trees, moreover, are stunted. Some of them may stand twenty feet tall or more, though their thick trunks taper rapidly to the tips. But in more exposed areas the trees are only three or four feet high, even though they may be hundreds of years old. In addition, clumps of shrubs, mainly willows, grow in river valleys, and the valley sides are

43

clothed in dwarf birches and low creeping shrubs such as cranberry, bilberry, bearberry, and crowberry. As you continue north, you finally reach a point where the trees disappear altogether. There, stretching on as far as the eye can see, are the immense, gently rolling, treeless plains sometimes called the "arctic prairie" but better known as *tundra.* From a distance the tundra looks bleak and rather monotonous, but actually it includes quite a variety of habitats. The treeless arctic prairies, in fact, include so many different plant communities that some botanists question the validity of lumping them all under the single catchall term, tundra. But for the sake of simplicity, we will stick to that one well-known word.

Over great areas, you will notice that the land is covered with a peculiar soil made up of coarse gravel and small lichen-covered rocks. This material, known as *till,* was deposited by retreating glaciers; as the ice sheets melted, they dropped the loads of rocks, sand, and gravel that were mixed up with the ice. Even on this poor soil there is a thin cover of lichens, peat moss, and scattered shoots of sedges.

Like natural powder puffs, the fluffy flowering heads of cotton grass sway with every breeze. Several varieties of these handsome perennial sedges thrive in marshes and other moist areas of the tundra.

Low, woody shrubs, like those that grow along valley sides, and some tiny dwarf willows also grow in protected crevices. All in all, however, plants cover only about one third of the ground. The rest is rock, bare soil, or dead litter, although in some areas the ground is almost completely covered with vegetation. Some of the depressions in this rolling landscape hold an entirely different kind of habitat. Here you may find a pond ringed by marshy soil where sedges, such as arctic cotton, grow in tussocks that project from beds of peat moss.

Arctic terns are likely to be found near the ponds, as are sandpipers of various kinds, and in the water you may see a little northern phalarope. This curious shorebird has a habit of spinning about on the water like a top, a maneuver that stirs up insects and other small animals from the bottom and brings them within reach of the phalarope's probing bill. Phalaropes are also unusual in that, unlike most birds, the females are more colorful than the males. Following mating, moreover, the gaudy female lays her eggs and then abandons them to her drab mate for the chores of incubation.

Another kind of habitat you will find in the tundra is

Among the throngs of birds that breed in the far north each summer are arctic terns. They arrive at their breeding grounds by mid-June, raise their young, and depart in late August and early September.

45

Near Campbell Lake, to the east of Canada's Great Slave Lake, eskers snake across a watery landscape. In the far north these well-drained meandering mounds of gravel and other glacial debris provide lookout posts and denning sites for wolves and arctic foxes. They were formed when silt and gravel on the beds of rivers running under glaciers were left high and dry as the glaciers melted.

long, low ridges of sand, silt, and gravel, called *eskers*. Some of them run on for miles, twisting and turning like rivers, and in fact they do mark the courses of long-extinct rivers that once flowed beneath glacial ice. When the ice finally melted, the silt and gravel deposited on the river bottoms remained in the form of ridges snaking across the tundra.

Eskers are especially important to two tundra animals: wolves and arctic foxes. Because of their height, slope, and porous nature, eskers are well drained. In addition, they are composed of materials in which a tunnel can be dug without immediately collapsing. As a result, both wolves and arctic foxes depend on eskers for denning sites in a land where soil suitable for burrowing is scarce.

No trees at all

The most striking characteristic of tundra is the total absence of trees. Why, you may wonder, are trees unable to grow there? Most botanists agree that the controlling factor is temperature. Winter temperatures are not too important since trees are dormant then, and once they are dormant, it matters little how cold it gets. But in summer when they are growing, plants must manufacture large amounts of food. The speed at which chemical reactions take place in all living cells is closely related to temperature, however, and apparently the cells of trees cannot operate fast enough to maintain their bulk when the temperature falls below 50 degrees. Although there are many exceptions due to local variations in growing conditions, in general, tree line is found where the average temperature of the warmest month, July, is 50 degrees.

Winter wind also is important. As it blows, it continually shifts the cover of snow, sweeping ridges and hilltops bare and depositing the snow into drifts in ravines and on the sheltered sides of obstructions. In addition, the wind carries sand and fine bits of gravel. This constant blasting by snow and sand is enough to kill the buds and growing tips of any trees that project above the snow cover. It is for this reason that the dwarf trees and shrubs of tree line are confined to sheltered ravines and other places where snow accumulates in winter. Buried beneath the snow, the growing tips are protected from the blasting wind.

Lack of moisture probably further limits the growth of trees. This may seem a little hard to believe at first, since the tundra is dotted with ponds and many areas are so wet and muddy that it is a good idea to wear boots while exploring. Yet the total annual precipitation seldom exceeds ten inches. In a hot climate, such meager rainfall would produce desert or near desert conditions. How can this seeming contradiction be explained?

Frozen soil

For one thing, the surface water evaporates very slowly because of low summer temperatures. You will find an even more important reason if you try to dig a hole in the ground. For the first few inches the digging is easy, but at a depth of a foot or so, you will reach a hard, rocklike layer. It is not rock at all, however; it is frozen soil.

Look closely, for this is your first glimpse of one of the most important phenomena of the far north: *permafrost.* In regions where the average temperature for the whole year is below freezing, the ground remains frozen except for a

layer near the surface that thaws each summer. Permafrost is the permanently frozen subsurface material, while the surface layer that thaws annually is called the *active layer*. The level marking the maximum depth of the summer thaw is the *permafrost table*.

Obviously water cannot penetrate the permafrost. Since the permafrost table is so close to the surface in the tundra, all the rain and meltwater from snow are held in the relatively shallow active layer, where they are more or less available to plant roots.

As you might expect, the far north is ringed by a continuous belt of permafrost. From the records kept by drillers of oil wells, we know that in some places the permafrost is as much as two thousand feet thick. However, it is not that thick everywhere, and in general it becomes thinner toward the south. In addition, the permafrost table tends to lie farther underground toward the south. Finally, we should note that at more southerly latitudes, there is a circumpolar zone where the permafrost is no longer continuous. In this zone of discontinuous permafrost, rather large areas are free of permanently frozen ground. Still farther south the permafrost disappears altogether.

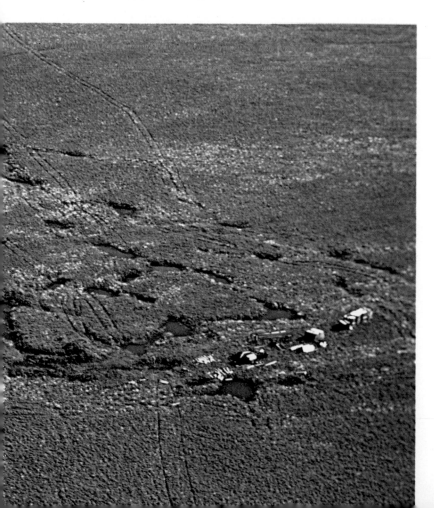

The work of oil drillers provides evidence of the fragility of the tundra landscape. Though this well is now abandoned, the land remains scarred by tracks of the drillers' vehicles, which crushed and tore up the vegetation and permitted the permafrost to thaw. As the pace of oil exploration accelerates in the far north, conservationists are demanding that steps be taken to minimize danger to the habitat.

As a matter of fact, we have been traveling over permanently frozen ground for quite some time. In western Canada, the southern limit of discontinuous permafrost lies slightly to the south of Wood Buffalo National Park. But the permafrost table there was so far underground that we were unaware of its existence.

Life beyond tree line

The dramatic change in habitat from boreal woodland to treeless tundra has profound effects on animal life. Animals in the boreal woodland are pretty much the same as those found in true taiga. As you approach tree line, however, animals that depend on trees for nesting, food, or escape from enemies gradually disappear from the scene. Fifteen species of wood warblers, for example, breed in Wood Buffalo Park, and at least twelve of them range north into the boreal woodlands. But only a few kinds are found in the scattered spruce groves at tree line, and none at all nest on the open tundra.

Throughout the high arctic, snow buntings build their well-concealed nests in rock crevices or among grass and moss on the open tundra. The five or six young, which hatch in about two weeks, are able to fly by the time they are twelve days old.

The jaunty black and white striped blackpoll warbler is one that depends on the stunted spruces at tree line. In the United States, blackpolls are a common sight during their spring and fall migrations from and to their wintering grounds in northern South America. During their long summer absence, they are up near the tree line in the far north rearing families, although some also nest at tree line on mountaintops in New England. Tree sparrows, which winter as far south as South Carolina and New Mexico, also head for tree line for the nesting season, and some go even farther north and nest out on the open tundra.

On the other hand, there is a whole array of ground-nesting and water-loving birds that nest on the tundra and rarely venture south of tree line during the breeding season. Snow buntings build their grass and moss nests among stones and in rock crevices, while Lapland longspurs hide their nests in grass hummocks or beneath dwarf shrubs. All sorts of plovers and sandpipers also migrate to the tundra for breeding, as do many kinds of ducks and geese.

Snowy owls are year-round residents of the tundra, depending primarily on lemmings and other small rodents for food both winter and summer. In winters when their food is scarce, however, they often wander beyond tree line and occasionally stray south as far as Georgia and Texas.

Very few birds live on the tundra throughout the year. The most impressive year-round residents are snowy owls, handsome white birds barred with dark brown streaks. They are common throughout the circumpolar tundra. Occasionally, in winters when their food is scarce, some of them stray as far south as the Gulf Coast, but otherwise they normally stay on the tundra year-round. Two kinds of ptarmigan also live on the tundra, feathered in browns and tans in summer and disguised in white plumage in winter. Some of them wander south in winter, passing even beyond tree line, but many remain on the snowy, treeless tundra throughout the months of darkness.

Mammal residents also change at tree line. A few hardy adaptable species live both in the taiga and on the tundra. The short-tailed weasel, commonly called ermine, and the wolverine flourish in both habitats. Wolves also are at home in both places, and red foxes and moose sometimes venture beyond tree line. Caribou, on the other hand, summer on the tundra and migrate to the taiga in winter.

Once you go beyond tree line, a great many mammals

drop out of the picture. Red squirrels, flying squirrels, and porcupines all are absent from the tundra. Martens, mink, fishers, and many others also disappear from the scene at tree line; but in their places, you will begin to see a number of animals that live only on the tundra. Instead of the snowshoe hares of the taiga, you will find groups of larger, much heavier arctic hares. Black bears do not range beyond tree line, but out on the tundra there are large, lumbering grizzly bears. Although grizzlies also live in coniferous forests on the western mountains, for some reason they are absent from the similar forests of the taiga.

Near rivers you will almost certainly see arctic ground squirrels. They are able to burrow here because the soil near rivers and lakes is free of permafrost. The large volume of water in rivers and lakes warms the nearby soil enough to prevent formation of permafrost or at least keep the permafrost table quite deeply buried. Ground squirrels also find denning sites among piles of broken rocks and boulders and in well-drained sandy glacial deposits.

Along rivers and in sedgy marshes you will find signs of

The grizzly bear, distinguished by the conspicuous hump on its shoulders, lives on the tundra in Canada and Alaska, but is also found in the mountain forests of British Columbia and the Rocky Mountain states. Reaching full size at eight or ten years, mature males normally weigh five to six hundred pounds while females average about four hundred pounds.

mouselike animals. Beneath the tall sedges are well-worn runways where the grasses and sedges have been cut away to form convenient but nicely hidden routes through the vegetation. Some of the trail networks probably were formed by meadow mice, or voles, while others are the work of closely related animals called lemmings.

Two kinds of lemming live on southern tundras. Both are stocky little volelike animals, five or six inches long, with very short tails and such dense fur on their heads that their ears are almost hidden. Brown lemmings have no prominent distinguishing marks, but their grayish heads and slightly reddish backs and rumps are colors that blend well with the bare soil, mosses, and lichens of their habitat. The slightly larger collared lemmings, in contrast, have a well-defined black stripe running down the center of their grayish backs. Collared lemmings are also called varying lemmings, since they become pure white in winter. Both species are found in the same areas and even use the same runways, but collared lemmings have a greater tendency to live on drier upland tundra as well as in marshy areas.

A muskox peers from a willow thicket, one of its favorite feeding areas in summer. The bag beneath each of the bull's eyes contains a gland which gives off a musky odor when rubbed against brush or the insides of its forelegs.

The beleaguered muskox

A good place to see another characteristic tundra animal is in the Thelon Game Sanctuary, about two hundred miles northeast of Great Slave Lake. The fifteen-thousand-square-mile sanctuary is bisected by the Thelon River which flows into Hudson Bay. In summer you may glimpse strange cowlike animals among the willows along the river. These are muskoxen, once common in many tundra areas but now quite rare.

They are impressive looking animals. Bulls are about 4½ feet high at the shoulder and may weigh as much as nine hundred pounds, although cows are quite a bit smaller. Both sexes are crowned by pairs of horns that almost touch at the bases and curve down along the sides of the head, then sweep up in sharp points at the tips. The most distinctive features of muskoxen are their coats of long, dark, flowing hair that sweeps down almost to the ground, making them look a bit like animated mops. In fact, attempts are being made to domesticate muskoxen for the sake of their

When threatened, muskoxen characteristically form a protective circle around their young and prepare to meet the danger head-on. Though effective against natural enemies, the technique makes the animals such easy targets that hunters in search of meat and hides at one time nearly exterminated Canada's muskoxen.

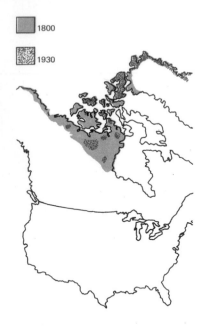

At the beginning of the nineteenth century, muskoxen ranged over most of the tundra in western North America, on many arctic islands, and along the northern coast of Greenland. By 1930, however, hunters had practically wiped them out in all but a small part of their former range. Since 1930, when muskoxen were given legal protection, they have increased slightly in number and are once again extending their range.

impressive coats. The extremely fine, dense hair beneath their outer coats, which keeps the animals warm throughout the arctic winter, can be spun into a fine high-quality wool. Proponents of the idea see domestication of muskoxen as a possible way of bolstering the income of Eskimos as they shift from ancient to more modern ways of life.

Although no one knows how many muskoxen originally lived in northern Canada, including the islands to the north of the mainland, we do know that at one point they were practically exterminated. Between 1862 and 1916, the Hudson's Bay Company alone bought at least fifteen thousand muskox skins, and no one knows how many more were sold to independent traders. Many of the animals were also slaughtered for meat. In the last century and early in this one, when arctic exploration was at its height, great numbers of muskoxen were killed for meat, both for the explorers and for the dogs that pulled their sleds. Still more were killed around the turn of the century by whalers to feed their crews.

Finally, muskoxen were popular in zoos, and large prices were paid for live calves. Capturing them was a simple but gruesome matter. In their home ranges, muskoxen have evolved a characteristic defense against their only enemies, arctic wolves. When wolves appear, muskoxen do not flee. Instead, the herd gathers into a circle with the bulls on the outside and the cows and calves at the center. Daunted by the wall of bulls brandishing their polished horns, the wolves generally leave the herds alone.

Observing this behavior, the calf hunters quickly learned to take advantage of it. When they found a herd, they released their sled dogs, causing the muskoxen to form a defensive ring as they would for a pack of wolves. From a safe distance the men then shot all the adults, leaving the calves helpless, confused, and easy to capture. Thus, to supply a single calf for a zoo, an entire herd of fifteen to thirty muskoxen often was destroyed.

As a result of all these forces, muskoxen were completely eliminated from many parts of their former range by the turn of the century. Only a few survived along the Thelon River and elsewhere in the arctic. In 1917, however, they finally were given complete legal protection. Ten years later, when the remnant herd was discovered in the Thelon Valley, the area was set aside as a game sanctuary.

Their salvation probably came just in the nick of time, but these measures seem to have been successful. In 1930, only five hundred muskoxen were thought to survive in all mainland Canada. Now there are about 1500 on the mainland, and they are spreading to other regions, while about 8500 muskoxen now thrive on remote, seldom-visited arctic islands. In addition, about 575 muskoxen—the offspring of stock brought from Greenland in 1931—live on Nunivak Island off the coast of Alaska.

North and south with caribou

Another characteristic animal of the tundra is the barren ground caribou, although many of them return each fall to the taiga where there are fewer winter storms and food is easier to obtain because the snow is softer. The closely related woodland caribou, in contrast, remains in the taiga throughout the year. The Old World also has caribou, including the semidomesticated variety commonly known as reindeer.

Your first view of a caribou may come as a surprise, since they are not as large as many people imagine. A full grown

Born in June, while snow still blankets much of the tundra, the frisky, ten- to fifteen-pound calf of the barren ground caribou is able to stand in about twenty minutes and can keep up with the cow within a few hours of birth.

bull rarely is more than four feet high at the shoulder. However, their magnificent branched antlers, sweeping back and up over their heads, give them an air of elegance. Unlike other North American deer, even the females have antlers, though they are smaller and less elaborate than those of the males.

In April or May caribou wintering in the taiga begin to make their way toward the tundra, gathering into seemingly endless herds that sometimes number in the thousands. It was these large herds that used to lure Eskimos inland from their usual haunts along the coast in order to hunt for easily taken supplies of meat and hides.

As they move out on the tundra, the caribou vary their staple diet of lichens with the fresh green growth of sedges. Later they begin to eat the tender new leaves of dwarf birch, willow, and other low shrubs, as well as a variety of tender plants such as louseworts, lupines, and avens. Later in the summer they also eat the fungi that spring up on the tundra.

Notable migrators, caribou stream across the landscape in herds that sometimes number in the thousands. After wintering in the taiga, they head for the tundra—sometimes to areas hundreds of miles away—and then in autumn they gradually drift back toward tree line.

The calves are born in June, even though there usually is still snow on the ground and the temperature is still below freezing. But if they can find shelter from the cold wind—beside their mothers or even behind boulders—most of the calves normally survive. Many are able to stand on their wobbly legs within twenty minutes of birth, and soon they are suckling their mothers for their first warm drink of milk. By the time they are a day old, the calves are strong enough to outrun a man if necessary, and soon are able to keep up with the migrating herds, whose summer wanderings may take them as much as eight hundred miles from their wintering grounds.

As summer wanes, the herds begin to drift back toward tree line. There the animals mate in October or early November. As winter tightens its grip on the landscape, they gradually move farther into the woods, roaming over large areas to find enough lichens to eat. But when days begin to lengthen again in spring, they will head north once more toward their calving grounds to begin another annual cycle.

ARCTIC NATIONAL WILDLIFE RANGE

Caribou (*left*) and polar bears (*below*) are
spectacular examples of the abundant wildlife that
finds sanctuary in Arctic National Wildlife Range.
This 8,900,000-acre preserve in the northeastern
corner of Alaska, created in 1960, serves as an
ideal natural laboratory where scientists can study
the ecology of the far north. Now that vast reserves
of oil have been discovered on Alaska's nearby
North Slope, threatening the region with
uncontrolled development, the need for refuges
where wildlife can be studied in an undisturbed
state becomes even more acute. Conservationists are
now urging the Canadian government to set aside
an adjacent area in the Yukon Territory to create
an international refuge.

Like frosting on a cake, huge permanent icecaps crown the mountain tops on the interior of Ellesmere Island, some six hundred miles from the North Pole.

A lonesome landscape

The next stage in our journey to the far north carries us far beyond the Arctic Circle to Lake Hazen, near the northern tip of Ellesmere Island. Here, off the coast of Greenland, we are almost at the northernmost limit of land and within six hundred miles of the North Pole.

The landscape you see as you fly in to the small research station beside the lake is bleak and rather forbidding. The fifty-mile-long lake is flanked by two mountain ranges, and both are topped by huge permanent ice caps. Sweeping down from the mountains are many glaciers, some reaching to within five miles of the lake. Although it is July, the lake is still almost completely covered with ice.

Clearly this is very cold country. It is no wonder the landscape seems almost totally devoid of plant life. Patches of green tint marshy areas, but otherwise the ground seems barren and desertlike. Nor is there much evidence of animal life. Near some of the marshes you may see a few musk-oxen or caribou, and perhaps a white wolf. A few greater snow geese, large white birds with black-tipped wings, may also be visible about their nests near the lake. But from the air you can detect few other animals.

Even the look of the land itself hints of a harsh climate. Steeper slopes are gouged here and there with huge mud slides. On more gradual slopes great lobes of soil point downhill between long lines of stony ground. Both the mud slides and the lobes result when wet soil of the active layer breaks loose from the permafrost table and slides downhill. This process is known as *solifluction* (literally, "soil flow"), and the mud lobes are called solifluction lobes. The strips of rocks between the lobes are due to forces set up by the expansion and contraction of soil in the active layer as it freezes and thaws, which pushes rocks to the surface and then sideways for short distances.

Because of the thin plant cover, you can easily detect another effect of frost action. In a marsh near the research station, the ground is broken up into many-sided more or less circular islands, low and water-filled at their centers and higher at the sides, with water-filled troughs running between. These *ice-wedge polygons*, a common sight on flat

The desolate landscape of Ellesmere Island, like much of the northernmost tundra, is patterned with geometric designs that result from repeated expansion and contraction of surface soil as it freezes and thaws.

lowlands of the tundra, apparently result from contraction of the ground when it cools. This causes cracks to form in the permafrost, much like the cracks that form in mud or clay when it dries. Water then seeps into the cracks and, when it freezes, expands and further enlarges the cracks so that more water can seep in. Eventually the network of cracks is filled with large wedge-shaped veins of ice, which, through their expansion, force the soil on either side up into ridges. Although the shallow centers of the resulting polygonal "islands" are not always filled with pools of water, the patterned surface of the ground resulting from this kind of frost action is always quite conspicuous.

Cold deserts of the high arctic

When you finally land and step from the plane, your earlier suspicions are confirmed. This *is* a cold place. Although it is midsummer, you probably will need a sweater or light parka. Even in July, the warmest month, the average temperature is only about 42 degrees. In fact, it is only in June, July, and August that the average temperature creeps above freezing, and it remains above zero for only five months of the year.

Just as summers here are short and cool, winters are long and very cold. Once the temperature drops below freezing in September, it remains there until late May or early June. The lowest temperature ever recorded at Lake Hazen was 70 degrees below zero, but even the average monthly temperature from December to March is colder than 30 degrees below zero.

Besides being a virtual deep-freeze in winter, this land far beyond the Arctic Circle is also a place of total darkness, while in summer the sun never sets. At the spring equinox, about March 21, the day is twelve hours long. For the next three weeks, day length increases fifty minutes per day until, by April 11, the sun ceases to set. For the next 143 days, the sun remains above the horizon for twenty-four

Patterned ground takes many forms in the Arctic. Here, on Bathurst Island, a honeycomb of dikelike ridges encloses pools of water about two feet in diameter. In another type of patterned ground, water-filled trenches encircle polygonal islands of dry land.

Among the strange land forms of the tundra are large pudding-shaped hills with cores of solid ice called pingos. This one is on the MacKenzie River Delta in northwestern Canada, where pingos are especially common. Although their origin is uncertain, one theory suggests that pingos are formed when shallow lakes are drained or dry up. As permafrost closes in beneath the old lake bed, the core of ice and covering soil are forced upward, just as the top pops from a bottle of milk when it freezes.

hours a day. Then in September, over the course of three weeks, the days again suddenly shorten from twenty-four hours to twelve hours. By mid-October, the sun disappears altogether and does not reappear until early March.

Thus, most arctic plants never see spring or fall. By the time they emerge from under the snow, the days already are twenty-four hours long, and when the sun sets, snow again blankets the northern landscape. To some extent, the twenty-four hours of daily sunlight in summer compensate for the shortness of the growing season. But living conditions still are difficult.

Wind is less of a hindrance to the survival of plants. Strangely, winds in these high arctic areas are more moderate than farther south. Average wind speeds at Hazen Lake are less than five miles per hour, and in winter there is frequently no wind at all. Winds strong enough to pick up and carry particles of ice and snow do occur, but they are rare and are more common in summer than in winter.

The final and perhaps most severe impediment to plant growth in the high arctic is lack of moisture. In many areas, the total precipitation is less than six inches, mainly in the form of snow. At Lake Hazen annual precipitation is only about one inch. The high arctic thus is often referred to as a "cold desert"—and with good reason.

Problems for plants

From ground level it becomes obvious that this far-northern outpost is not as desolate as it seemed from the air. The vegetation is not exactly lush, but most of the land has at least a sparse cover of plant life.

The vegetation is not particularly varied, however. There are plenty of mosses and lichens, of course, but there are relatively few kinds of the so-called "higher plants" (root-bearing, seed-producing plants). As a matter of fact, botanists have discovered only 115 species growing around Lake Hazen, while on nearby Ellef Ringnes Island only 49 species have been found. On Greenland, on the other hand, there are about 450 species of higher plants, and about 650 species occur on the tundras of mainland Canada and northern Alaska. By way of contrast, the tiny British Isles can boast a flora including thirteen thousand species of higher plants.

Obviously not many plants have the means to cope with the harsh living conditions of the high arctic. The most severe limitation to their growth, most botanists agree, is lack of moisture rather than low temperatures. If you look around, you can see why. Very few plants grow on dry hill-sides or in well-drained sand and gravel soils. Low, wet

Winter in the high arctic creates problems for man and beast alike. However, while humans can build shelters and burn fuel to keep warm, wild animals must depend on their natural adaptations to survive the frigid blasts of an arctic blizzard.

areas and places where springs reach the surface on clay slopes, in contrast, are quite thickly vegetated.

Nutrients such as nitrates and phosphates also are in short supply on the uplands. The seepage area below the garbage pit near camp, for example, is noticeably greener than its surroundings; the garbage is, in effect, fertilizing the ground. The same is true of the area around a nearby arctic fox's den. The animal's droppings have enriched the soil, permitting a lusher growth of plants. Even on dry uplands a few orange and green patches stand out in contrast to their bare surroundings. Here again, crustose lichens on the rocks and cushions of willow and cinquefoil between them have been fertilized by droppings around the perch of a snowy owl, a long-tailed jaeger, or a rock ptarmigan.

To survive, however, arctic plants must cope with far more than shortage of water and of nutrients. They also must be able to carry on photosynthesis and other life activities at temperatures near freezing. Their roots must grow practically in contact with permafrost and yet not be destroyed by frost action in the soil. They must be able to withstand freezing at any stage in their development and, when they thaw, be able to pick up where they left off. Flowers, for example, may be encased in ice or buried under snow for several days and then thaw out and eventually produce healthy seeds. Finally, they must be able to function at low light intensities and take every possible advantage of the short growing season. To get a head start on the season, some species actually begin to develop while still covered with a blanket of snow in spring. Others store huge reserves of fats and carbohydrates in their roots before becoming dormant in the fall, and then draw on the reserves to get an early start when they begin growing in spring.

All in all, living conditions are so severe that only highly specialized plants are able to cope with them. It is no wonder that so few species have mastered the rigorous challenge of survival in the high arctic.

Minerals seeping from a stranded whale bone provide enough essential nutrients to support the growth of a colorful clump of arctic poppies and other plants. In many areas of the arctic, shortage of nutrients is one of the main factors limiting the growth of plants.

Neat tufts of cinquefoil dot nearly barren, gravelly areas of tundra where few other plants can survive. Many kinds of cinquefoils grow in various tundra habitats.

71

Just as plant life at Lake Hazen is quite limited in variety, animals are much less varied than in more temperate regions. Judging from the number of flies and mosquitoes, for instance, you might imagine that you were in an insect paradise. Yet only about three hundred species are known to occur in the area, and more than half of them are flies of one sort or another. Here and there, you may spot a bright-orange and black *Boloria* butterfly fluttering about or simply basking in the sun; in all, nineteen species of moths and butterflies live around Lake Hazen. But most of the insects are less conspicuous, living close to the ground or in the soil itself.

Birds are much easier to find. On the margins of ponds you are certain to see a few red-legged birds strikingly patterned with black, white, and reddish-brown. The birds are ruddy turnstones, named for their habit of using their bills as pries to turn over rocks in order to get at animals hiding underneath. Probably the commonest birds at Lake Hazen, the turnstones are especially interesting because they belong to the European rather than the North American race of their species. Instead of wintering in the United States and South America, they winter in Africa and migrate back and forth across Europe and Greenland.

Other common waders along pond margins are knots, chunky birds with mottled backs and cinnamon-colored undersides. Like the turnstones, the Lake Hazen knots come from Africa by way of Europe. And like the turnstones, they bring parasites with them, mainly tapeworms and mites. However, their parasites are more like those found on other African shore birds than those found on knots and turnstones living elsewhere in North America.

On uplands away from water two other fairly common birds are snow buntings and hoary redpolls. The redpolls, sparrowlike birds with bright red caps, return to Lake Hazen and begin nesting while the landscape is still in the grip of winter. Young birds have been seen flying as early as June 27, and by the end of July most of them have departed for their wintering areas.

The sum total of birds found at Lake Hazen is a mere twenty species, but they provide an exceptionally good

The ruddy turnstone finds food by flipping over rocks, shells, and other objects with its bill in order to get at small animals hiding underneath. If the stone is a large one, the bird may complete the job by pushing it over with its breast.

Every summer, ruddy
turnstones converge on the
coasts and islands of the
high arctic, returning from
wintering areas as far away
as Africa and South America.
Their eggs, usually four to
the clutch, are incubated in
shallow nests on the open
tundra or tucked in crevices
among rocks.

THE IMPERILED PEREGRINE

The thrilling sight of the peregrine falcon, our swiftest bird of prey, soon may be a thing of the past. North American peregrine populations have ebbed to such a low level that the species is in danger of extinction.

The animals of the arctic soon may number one less. The peregrine falcon, or duck hawk, once widely distributed across North America, now is virtually extinct except in Alaska and northern Canada. The growth of urban areas and disturbances at nesting sites have played a role in the majestic bird's demise, but the main culprits seem to be DDT and other chemicals. When the bird eats prey

that has fed on insects that consumed the chemicals, residues of the poisons enter the falcon's body, apparently upsetting its hormonal balance. The result is sterile eggs or eggs with shells so thin that they break before hatching. Even birds that nest in the far north are not safe from the pesticide menace, for they consume the deadly chemicals with their prey when they migrate south in winter.

On its rocky eyrie on a cliff overlooking the Anderson River in Northern Canada, a peregrine falcon tends to her recently hatched chicks. If the young birds survive, they will require two years to grow to maturity.

Although arctic hares on Ellesmere Island and in northern Greenland remain white year-round, in most areas they shed their white winter coats for gray or light-brown ones in summer....

demonstration of circumpolar distribution. Eighteen of them are also found on the opposite side of the globe in Eurasia. The other two have Eurasian counterparts that are so similar that they also may ultimately prove to be merely varieties of the species found at Lake Hazen.

The high arctic mammals found here also are circumpolar species. In every case but one, exactly the same species occur in northern Eurasia, although one of them, the musk-ox, became extinct in Eurasia a few thousand years ago. The only possible exception is the arctic hare. A kind of hare inhabits extreme northern Russia, but it does not extend to the islands off Russia's north coast, so it may not be exactly the same as the arctic hare found on Ellesmere Island.

The truly astonishing thing about the mammals here,

however, is that there are only seven species. Four of them are *herbivores,* or planteaters: collared lemmings, arctic hares, caribou, and muskoxen. The other three are predators: ermines, arctic foxes, and wolves. And that is all there are.

Even so, it is intriguing to observe the relationships among them. The predators, for instance, neatly correspond in size to the mammalian herbivores. The little ermines probably are capable of killing only lemmings, while the medium-sized arctic foxes can handle both lemmings and arctic hares. The large wolves, in turn, are powerful enough to pull down both caribou and muskoxen.

Observations of this sort are especially meaningful to *ecologists,* scientists who study the interrelationships of living things with each other and with their nonliving environ-

... The sociable hares, which may weigh as much as twelve pounds, live on windswept, rocky slopes and tundra uplands. Occasionally they surprise observers by hopping upright like kangaroos.

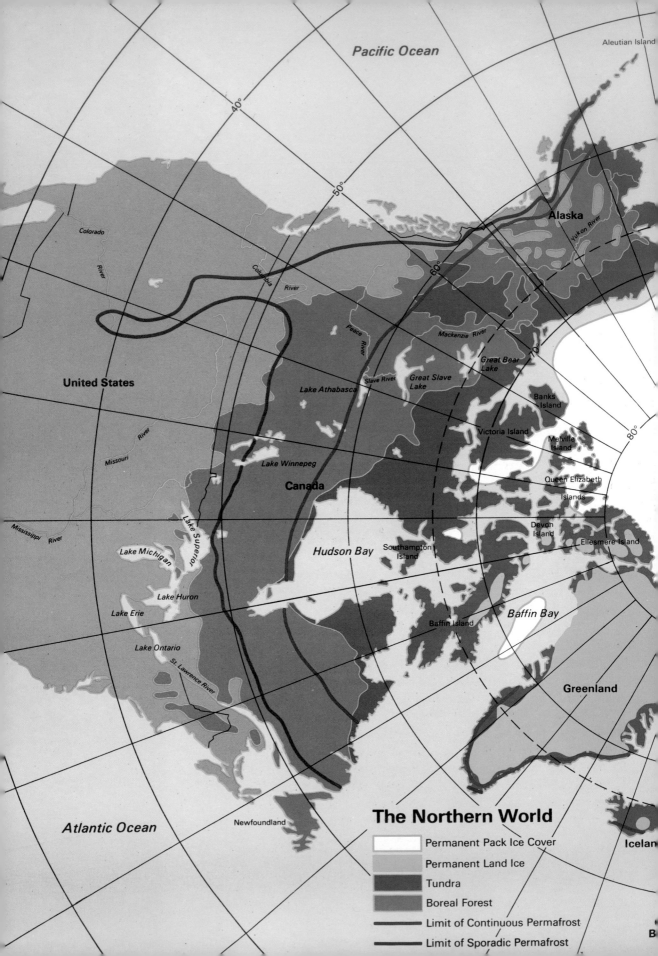

The Northern World

	Permanent Pack Ice Cover
	Permanent Land Ice
	Tundra
	Boreal Forest
——	Limit of Continuous Permafrost
——	Limit of Sporadic Permafrost

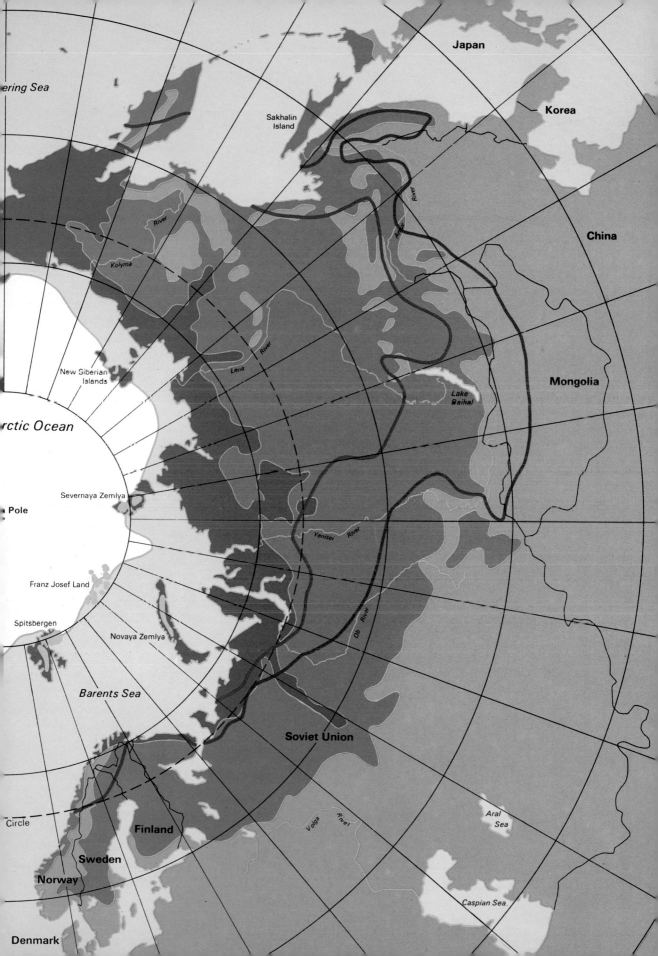

ment. In fact, for ecologists it is the sparsity of life forms that makes the whole far north a unique living laboratory. It is much easier to understand the relationships among plants and animals in a place where there are relatively few life forms than it is in more populous environments. By learning how things work—and what happens when things go wrong—in this relatively simple environment, they hope to find clues to the functioning of more complex ecological systems.

Land's end

Beyond Ellesmere Island, the far north changes radically in character, for here we are at land's end. Off the northern coasts of North America, Eurasia, and the northern islands, there is a huge dent in the top of the globe, and the dent is filled with the Arctic Ocean. This, however, is an ocean unlike any other in the world: most of its surface is covered throughout the year by a heaving, grinding sheet of ice. Even in late summer when the ice cover has shrunk to its minimum, the permanent polar ice pack is ringed only by relatively narrow channels of open water. In midwinter, the ice extends from shore to shore, sealing almost all the northern coasts with ice. Even in winter, however, the ice cover is not complete. The movement of ice by wind, tides, or currents opens up channels of frigid water.

Even in this strange environment there is life. As in any body of water, the primary source of life is minute drifting one-celled algae. Although algae grow slowly in these nutrient-poor waters, they grow best around the edges of the ice, where nutrients that slowly leak out of the frozen seawater are concentrated. Farther south, where Atlantic or Pacific waters meet those of the Arctic, currents bring nutrients from the bottom of the sea which enrich the cold waters. Here, algae grow abundantly.

Tiny animals eat the algae and in turn are eaten by fish and shellfish, which are themselves eaten by animals such

Throughout the year, most of the Arctic Ocean is covered by a permanent polar ice pack. Here some of the grinding sheets of ice have piled up in jumbled masses on an arctic coastline.

as seals and sea birds. Eskimos, too, have traditionally depended on the Arctic Ocean. Though their way of life is rapidly changing, some of them still cling to their ancient ways and turn to the sea for its meager but dependable harvest of fish, seals, and whales.

You would be lucky to see many of the whales that once migrated annually to arctic seas. The blue whale, for instance, once was fairly common along the edges of the arctic. Though it is the largest known animal—specimens one hundred feet long have been taken—it feeds mainly on tiny shrimplike creatures which it strains from the water with fringed comblike plates in its mouth. However, these magnificent creatures have been hunted so ruthlessly by commercial whalers that they are now all but extinct in the Arctic and are even becoming extremely rare in their last strongholds in the Antarctic. Much the same fate has befallen the bowhead, found only in arctic waters, and most

The icebreaking oil tanker *Manhattan* tears a path through the ice of Baffin Bay, passing the smaller icebreaker *Louis St. Laurent.* On its second trip to the arctic to prove the feasibility of shipping Alaskan oil by surface tanker, the giant *Manhattan* was gouged twice by ice and its hull cracked slightly—even though it had been strengthened after its first trip—thus dramatizing the possibility of suffocating oil slicks on arctic waters if such tankers are put into operation.

of the other whales that used to migrate to the edge of the arctic seas.

A few are holding their own. White whales, or belugas, still swim in small schools among openings in the drifting pack ice, their fifteen- to eighteen-foot-long white bodies shimmering like apparitions in the icy blue water. Mixed with the white whales there may be a few smaller, grayish narwhals. They are without any doubt the strangest-looking of all whales; projecting like a slender lance from the snout of adult males is a spiraled, ivorylike tusk as much as eight feet long. Actually a grotesquely overgrown tooth, the narwhal's tusk was thought during the Middle Ages to be the horn of the mythical unicorn. Eskimos still carve narwhal tusks into tools and decorative objects and hunt both narwhals and belugas for flesh, oil, and hide.

Narwhals are also hunted by killer whales, wolves of the sea which travel about in groups of as many as forty indi-

viduals. Once feared and persecuted by man, these handsome black and white dolphins with high triangular back fins are beginning to be appreciated and protected. Intelligent predators, they will kill young or injured whales or walruses, but are easily driven away by adults in good condition. They eat mostly porpoises or the smaller seals. Even seals that take refuge on an *ice floe* (a floating sheet of ice) are not necessarily safe from their attacks; killer whales have been known to nudge floes with their heads in order to dislodge potential meals.

Their choice of prey among seals is extremely varied, for many kinds of seals live in the Arctic Ocean. The most unusual are walruses, easily distinguished by their large size and pairs of long downward-projecting tusks. Walruses were long believed to use their tusks like clam rakes, diving to the ocean bottom and then probing with their tusks to dig up clams and other shellfish, but recent research suggests that they probably use their snouts to root out their food from the bottom, much as a pig roots out truffles.

Proud of his catch, an Eskimo boy strips the edible hide from his first whale, a beluga, or white whale. Though the Eskimo way of life is changing rapidly, the flesh of seals and whales still is the mainstay of the natives' diet in many villages.

In Alaskan waters, some of the commonest seals are northern fur seals, which during the breeding season gather in immense herds on the Pribilof Islands in the Bering Sea. Their thick, lustrous pelts have always been so highly valued for fur that, by the beginning of this century, they were practically exterminated. Since then, however, they have been protected by an international agreement, and the herds have again become large enough to permit carefully supervised harvesting for pelts.

Ribbon seals also live in Alaskan waters, the dark-brown males marked with whitish ribbonlike bands around their necks, rumps, and forefins. Ringed seals, the commonest and most widespread seals of the arctic, also have unusual markings. Their fur is mottled with white circles enclosing patches of black. Other wide-ranging seals of the Arctic include harp seals, harbor seals, bearded seals, and hooded seals.

All the seals are important to the economy of Eskimos, who hunt them for their skins, their meat, and for the thick

85

Practitioners of a vanishing
way of life, Eskimo hunters in
search of seals drive their dog
sled to the very edge of the
ice near Point Hope, Alaska.
In addition to consuming the
seals' flesh, they use the hides
for clothing, extract oil from
the blubber to heat and light
their homes, and carve the
bones and, in the case of
walruses, the tusks into
implements and ornamental
objects.

Popping to the surface for a breath of air, a harp seal unwittingly exposes itself to the hunter's club, harpoon, or rifle. In addition to the take by Eskimos, great numbers of these and other seals are harvested each year by commercial sealers. Their pelts are transformed into fur coats, their hides into fine leather, and their blubber into an oil used in creams and lotions.

In some areas, walruses are virtually the staff of life in the traditional Eskimo economy, with nearly every part of their bodies put to use. Now that Eskimos are hunting with rifles, however, walruses are declining in numbers in some places. Their ivory tusks are especially valued; when carved into trinkets, they can be sold for the cash needed to buy ammunition for the Eskimo's rifle.

Before they had rifles, Eskimos hunted seals with harpoons, waiting patiently to spear the animals when they surfaced at breathing holes in the ice. When the seal was harpooned, the weapon's barbed head separated from the shaft, but remained attached to a line on which the Eskimo played the seal until it was exhausted and could be hauled to the surface.

insulating layers of fat that help keep seals warm in the icy water of the Arctic Ocean. The main requirement for the Eskimos' traditional hunting method is an immense store of patience. Since seals are air-breathing mammals, they must surface every ten minutes or so to breathe. They make small breathing holes where the ice is thin. As long as they keep the holes from freezing over, they can stick just their noses above the surface from time to time to get a gulp of air. Aware of this habit, Eskimo hunters formerly had to wait patiently at a breathing hole, sometimes for hours at a stretch. When a seal finally pushed its snout through the hole to breathe, the Eskimo quickly thrust a harpoon through the hole to kill it. Now, with rifles, hunters can shoot the seals as they rest on the ice.

On polar ice

Out on the polar ice pack, seals are also preyed upon by the majestic polar bear, which still inhabits a range that extends all the way around the Arctic Ocean. Most polar bears appear to spend the greatest part of their lives right out on the ice. Recent studies along the shores of Hudson Bay, however, have shown that a substantial number of bears may go as far as one hundred miles inland during the summer. Females, of course, also frequently come to land and excavate dens in snowdrifts where they spend part of the winter and where they give birth to their young.

Studies by Russian and Norwegian scientists have shown that the main food of the polar bears that live in the northernmost parts of the range is seals, especially ringed seals. The Hudson Bay population, surprisingly, has a more varied diet. Animals summering along the west coast of the bay eat a tremendous variety of foods ranging from seals to muskrats and field mice; oldsquaw ducks; marine invertebrates; saltwater algae; rushes; sedges; and the leaves, stems, and berries of a dozen species of land plants. More than 80 per cent of the food of this population consists of plant remains and only 6 per cent seals. In contrast, the population summering around some islands in Hudson Bay depends neither on plants nor on seals for the bulk of its food. Instead, it specializes on sea birds. Oldsquaw ducks are of first importance, followed by eiders, brant, and other

With water sluicing from its fur, a polar bear prepares to mount an ice floe. This expert swimmer and diver moves through the water by paddling with its forepaws. Cubs sometimes hang on to the mother's rump or tail to get a free ride as she swims through the cold brine.

Equally at home on ice or in the water, the polar bear wanders from ice floe to ice floe in its endless search for seals and other food. A skillful hunter, it depends on stealth and camouflage as it creeps up on unsuspecting prey.

ducks, and gulls and loons, which together make up nearly 60 per cent of the bears' diet.

Polar bears do not attempt to capture seals in the water. Instead, they hunt them when the seals come out on the ice to rest. The seals are wary, however; they usually stay near the edges of ice floes or beside escape holes where they can plunge quickly into the water. Thus, despite the bears' massive bulk—adult males weigh between one thousand and fourteen hundred pounds—they must depend on stealth rather than strength to capture seals. To some extent the bears' whitish or slightly yellowish fur camouflages them as they creep across the ice and snow. Hiding behind chunks of ice, or flattening themselves to the surface so that they look like just another irregularity in the ice, they inch slowly forward until they are within striking distance. Then they make the final rush and clout the seal with one of their massive forepaws. One blow is usually enough, and the bear then settles down to enjoy its meal.

Sea birds, on the other hand, are probably captured while

sitting on the water. The bear dives some distance from a floating raft of birds, swims cautiously under the water until it is below its prey, and then floats up and grabs an unsuspecting bird from below.

As likely as not, however, a hunting bear will not be alone. A snow-white arctic fox often lurks nearby, curiously observing the bear's every move. Some arctic foxes spend the entire year on the tundra, but others move out to the coast or the ice pack in the fall when the ice forms. Here they can scavenge on dead seals and walruses or small animals washed up on the beach or frozen in the ice. Frequently they adopt a polar bear and tag along at a safe distance behind for weeks or even months at a stretch. Each time the bear kills a seal, its relentless camp follower moves in to feast on any scraps of meat that the bear leaves behind. When the bear moves on again, with the fox trotting along behind, the two of them are the only signs of life in the vast frozen world of the farthest north.

Until the airplane gave man with his high-powered rifles easy access to parts of the pack ice, the polar bears' existence was not seriously threatened. By 1965, however, it was

The snowy white arctic fox is a frequent consort of polar bears. Trotting along behind at a safe distance, it cleans up any scraps of meat the great hunter of the polar seas leaves behind when it makes a kill.

clear that overhunting in many parts of the bears' range was a real threat. In that year, concerned people from the United States, Canada, the U.S.S.R., Norway, and Denmark met in Fairbanks, Alaska, to take steps to ensure the bears' survival. New, more restrictive laws in several countries have reduced the annual kill, and an international team of scientists is studying the bears' movements, food habits, breeding, and survival under the coordination of the International Union for Conservation of Nature and Natural Resources. This is an outstanding example of international cooperation in the conservation of a species which can be saved only if all the northern nations act together.

Journey's end

In our travels through the far north we have come a long way since our beginning on the Peace River. As we journeyed from the spruce forests of the taiga, past tree line, and across the tundra to the polar ice pack, we have seen many changes along the way. From a land of dense forests we have come to a place of perpetual ice and snow. At the end of our trip, we have left the four seasons far behind and entered a strange realm where a six-month-long summer "day" alternates with an equally long winter "night."

Yet at each stage of our journey we have seen life of one sort or another. Although the number of kinds of plants and animals has diminished as living conditions became more and more rigorous, a few animals, as we have seen, manage to survive even on the polar ice. Now it is time to turn to an intriguing question: how does any life at all manage to survive the harsh winters and brief summers of the far north?

As generations of their ancestors have done, a group of Eskimos prepare to set out to hunt for whales from Point Hope at the northwestern tip of Alaska. Like the polar bear, the fox, and the prey they seek in arctic waters, the Eskimos are part of the life of the far north and have learned the secrets of survival in this extreme environment.

Land of

Ice and Snow

Perched on a limestone cliff overlooking Great Slave Lake is the Heart Lake Biological Station of the University of Alberta. The station was established as a place for studying what taiga mammals do in winter. Each year students from the university go there to live for several months and carry on scientific studies. If you could join them, you would find that the surrounding wilderness forms a superb natural laboratory. By keeping track of the signs of life you see, you would get a good idea of what kinds of animals spend winter in the taiga. You would soon begin to understand how they manage to survive in the far-northern land of ice and snow.

It is best to arrive in mid-September so that you can get your bearings before winter sets in. The laboratory is surrounded by a forest composed mainly of jack pines with an understory of junipers. (Jack pines dominate the Canadian taiga on well-drained sandy or gravelly soils that are too dry for spruces.) Nearby are stands of white spruce, aspen groves, mixed forests, and black-spruce bogs.

In September, the aspens and birches already color the landscape with patches of bright yellow, while the scattered willows have turned to russet brown. On the forest floor, smaller plants such as arctic bearberry form scarlet mats, and in the bogs golden larches stand out in warm contrast to the somber gray-green of black spruces. Larches are the only North American conifers that do not have evergreen needles. In August the needles begin to turn yellow and gradually rain down into the mosses. But in spring, clusters of fresh, new green needles will once again cover the dangling twigs of all the larches.

Autumn animals

As you explore the forest, you are certain to be impressed by the silence. By now most of the summer birds are gone, and the few that have not yet departed flit furtively through the branches. You might be lucky enough to see a stray gyrfalcon or a late flock of geese migrating south from the Arctic. Occasionally a gray jay glides away silently on outstretched wings, and now and then a boreal chickadee sounds a few weak notes. But most of the warblers, flycatchers, thrushes, and other birds that enlivened the scene in summer already have headed south to warmer climates.

This is hardly surprising. The woods, you have probably noticed with relief, are now quite free of insects. The hordes of mosquitoes that make life miserable in summer begin to thin out as the weather begins to cool in August. By late September, there are no more to be seen. If the day is warm, a few black flies probably are buzzing about your head, but soon they too will be gone. You might find a few ground beetles crawling through the moss and an occasional wasp or two. On the whole, however, insect food seems to have become quite scarce in the north woods.

Prospects for fruit-eating birds are not much better. Roses and certain kinds of cranberries and bearberries are still laden with fruit and will, in fact, keep their fruit right through winter. All the raspberries, gooseberries, strawberries, cloudberries, and most other fruiting plants, on the other hand, have been stripped practically clean. The few berries that were not eaten by bears and small birds have fallen to the ground and are disintegrating.

The straight, slender, cone-bearing larch, or tamarack, is common in cool damp areas of northern forests. Its inch-long needles, borne in rosettelike clusters along the branches, turn yellow and fall in the autumn, making the tamarack one of the few coniferous trees that is not evergreen.

Food is still plentiful for one bird, however—the needle-eating spruce grouse. As you walk along, one of them suddenly utters its alarm call from the branches of a spruce almost directly overhead and launches into flight. The furious beating of its short, rounded wings produces a sudden burst of sound as the bird hurtles between the branches of the trees, its body tilting this way and that. Then you notice several more spruce grouse sitting upright on the branches of nearby trees. They are calling softly to one another and pumping their necks up and down, apparently trying to decide whether or not to fly. But if they do, they will not fly far, for they are one of the few birds that remain in the taiga throughout the winter. In the coming months, you can look forward to seeing them often as you explore the forest.

Just ahead is the sign of another year-round resident of the taiga. Beneath the spruces is a great heap of brown scales from spruce cones, with a scattering of unopened cones on top. The heap is two or three feet high, twenty or thirty feet across, and riddled with holes. It is the *midden,* or refuse heap, of a red squirrel, built up over a period of years by a succession of squirrels that have come to this place to tear apart the cones and eat the seeds.

As you draw near, the present owner scolds with a loud, rattling chatter from the branches of a nearby spruce. He puts everything he has into the challenge, ending with a series of sharp chirps that seem to shake his entire body, from the tip of his nose to the end of his fluffy tail. If you take one more step, he will scurry up to a higher branch and repeat the message of his displeasure. Evidently he is too busy harvesting cones and storing them in tunnels in his midden to put up with intruders just now.

On the forest floor are clues to the activities of still other animals. Crisscrossing the ground are clearly marked trails four or five inches wide worn into the mosses and lichens. Pea-sized brown droppings in the trails identify the makers as snowshoe hares, which travel the same regular routes as they make their daily rounds. Squirrels also take advantage of the easy going provided by well-worn hare trails.

Scattered among the moss are a few mushrooms and other fungi. If you look closely, you will see that some are scarred with tiny marks made by pairs of long curving front teeth. Most likely they were made by red-backed voles,

The white spruce is one of the commonest trees of the north woods and one of the most majestic, sometimes reaching heights of over 150 feet. Its short blue-green needles, which emit a strong odor when crushed, have a whitish cast that accounts for the tree's name.

Snowshoe or varying hares are year-round residents of the taiga. Traveling along well-worn paths through the underbrush, they forage on bark, twigs, buds, grasses, and leaves. Their diet in winter is less varied than in summer and, in years when they are numerous, they sometimes nearly deplete their food supplies by the time spring arrives.

short-legged, stocky animals whose ears are nearly hidden in their long fur.

Deermice also live on the forest floor. Much sleeker than voles, their two-toned bodies are grayish-brown on top and nearly white below. Although their bodies are only two inches long, they have three-inch-long tails, in contrast to the red-back's scanty inch-long one. Large bulging coal-black eyes, prominent ears, and a slender pointed snout all contribute to the deerlike look which accounts for the deermouse's common name, although it is sometimes known as the white-footed mouse—again for obvious reasons. If you stop to reflect, you will realize that the long tail and those large ears may prove difficult to keep warm in winter. It will be interesting to see how the deermice and the stockier red-backed voles fare as the season progresses.

SUMMER

AUTUMN

Light show in the sky

Before turning in for the evening, you ought to step outside for a few minutes. The night is moonless and the forest is pitch black, but the sky is filled with stars shining with a clarity such as you probably never have seen through the polluted air of a southern city. There are no clouds to dull the brilliance of the stars or to shield the earth as it radiates its heat into outer space. From the chill in the air, you suspect that there may even be frost in the morning.

As you think about the prospect, a faint greenish glow suddenly appears in the northern sky. Seconds later a pale streak snakes across the heavens, then grows to a dancing curtain of bright green rays edged with the palest purple. The *northern lights*, or *aurora borealis*—for that is what they are—continue to dance briefly then die down almost as quickly as they came, only to reappear and light the sky in another direction.

As the eerie spectacle continues, the sky may be illuminated with blues, greens, purples, yellowish lines, and occasionally even sheets of red. Sometimes they appear as shimmering streaks, sometimes as waving curtains of color that resemble billowing sheets of silk, or even in cascading ripples like fantastic celestial waterfalls.

It is no wonder that northern Indians, confronted with such unearthly visions, constructed all sorts of legends about

WINTER

In summer, the snowshoe hare's brown coat blends well with the litter on the forest floor. In autumn, however, patches of white fur begin to replace the brown, and, by the time snow covers the ground in winter, its coat is completely white. In spring its white coat is replaced once again by a brown one.

In spring and fall, northern skies often shimmer with the aurora borealis.

Like many birds, snow geese escape the cold of the northern winter by migrating to warmer climates. After nesting on arctic coasts, they head south in September to spend the winter along the coastlines of the southern United States and Mexico.

the mysterious silent lights, imagining, for instance, that they could call them down to earth by rubbing together the abrasive strips on two matchboxes. We now know that the lights are caused by bombardment of the earth's outer atmosphere near both poles by charged particles radiating from the sun. When the atomic particles collide with rarefied gases at high altitudes, they cause the gases to glow, much as colored light is produced in a neon tube. The lights, most common in spring and fall, provide an almost nightly spectacle in the far north and occasionally are strong enough to be seen as far south as the southern United States. Recently it has been discovered that the auroras of the Antarctic occur at the same time as and, in fact, mirror the auroras of the north.

How to escape the cold

As September draws to a close, days get noticeably shorter, and nights grow longer. Auroras light the sky almost nightly. Temperatures continue to drop until, each morning, a skim of ice begins to form on puddles. Finally one day the

104

sun grows so weak that, by evening, the morning film of ice remains unmelted on larger puddles.

By now only an occasional migrating bird passes by the laboratory, and the woods seem more silent than ever. The birds, it seems, have hit upon the best solution of all for coping with the rigors of the coming winter—they simply go someplace else where it is warm. The change is most obvious farther north on the tundra. In summer the tundra is alive with ducks, geese, swans, gulls, shorebirds, and scores of others. In winter, snowy owls and ptarmigans are virtually the only ones that remain. As autumn approaches, the rest head for the United States, Mexico, South America, and other warmer climates. One of the champion migrators is the arctic tern. After breeding on arctic islands and far-northern coasts, it heads south to winter near Antarctica; when summer ends at the north pole, it seeks the coming summer at the opposite end of the globe.

Mammals, on the other hand, are less mobile than birds.

Grizzly bears escape the winter by denning up in sheltered lairs, where they lapse into a semidormant state resembling hibernation. While living on their stores of fat, their body temperatures remain near normal, although their breathing and heart rates slow down as they do in true hibernators.

HOW BEAVERS FACE THE WINTER

Beavers are one animal of the taiga that remains active throughout the year. By the time autumn colors have tinted the trees around their ponds (*left*), they are ready to meet winter head-on. Warm, waterproof fur, thick layers of fat, and stocky builds all help them resist the cold. And even when their ponds are sheathed with ice, the provident beavers have plentiful food stores near at hand. Throughout the summer, they fell trees with their chisellike teeth (*below*), then snip off the branches and drag them into the water, where the branches form a raft with some of the stems anchored in the muddy bottom. Then, when the icy grip of winter comes to the far north, the beavers swim from their fortresslike lodges to the sunken food stores, retrieve the branches, and gnaw on the nutritious bark.

A pair of five-day-old beaver kits nurses within the warm dry shelter of the lodge. The domed structure, with its floor above water level, is a snug retreat in winter. Constructed of sticks, mud, and debris, the plastered walls freeze to a rocklike shield that helps hold body warmth in while keeping wind and predators out.

The beaver, able to stay submerged for three minutes or more without breathing, is uniquely adapted for life in the water. Its webbed hind feet form efficient paddles, while transparent eyelids permit it to see underwater. Valved ear openings and nostrils close to keep water out, and flaps of skin that close behind its teeth enable it to gnaw bark while underwater.

Most of them remain in the same general area throughout the year, and only a few of them such as whales and caribou migrate. As we have already seen, many of the barren ground caribou leave the tundra in autumn to winter in the boreal woodlands. As they migrate south, the wolf packs that prey on them also move south to keep near the caribou.

A few mammals have evolved an entirely different means for escaping the winter cold. Earlier in the fall, you probably noticed a few woodchucks from time to time in more open areas near the laboratory. But now they are all gone. Covered with thick layers of fat, they have retreated to underground burrows where the temperature will not fall below freezing and gone into *hibernation*. Their normal body temperature of about 98 degrees has dropped to perhaps 36 degrees, and all their bodily activities, including breathing rate and heart rate, have slowed down drastically. In this state of greatly reduced animation, they use only a small fraction of the energy that would be required by an animal active at normal body temperatures.

In late fall you may occasionally notice least chipmunks darting among the shrubbery collecting seeds and berries to store in their networks of burrows. Chipmunks hibernate, but unlike woodchucks and jumping mice, they do not store much energy as fat. So, when they wake up from time to time during the winter (as all hibernators do) and their body temperatures return to normal, they eat some of their stores of food, then go back into hibernation.

Bears, in contrast, go into a state of dormancy. The black bears of the taiga and the grizzlies of the tundra den up in secure places in the fall and sleep through the winter. However, although their breathing and heart rates slow down, their body temperatures remain close to normal. They simply live off the energy stores in their fat. Female polar bears breed inland from the arctic coast and den up in autumn, but most of the males remain active throughout the winter. Like other bears, the female polar bears give birth to their cubs while in their winter dens.

The least chipmunk, like other chipmunks, spends the winter hibernating in its burrow. Like other hibernators, it wakes up from time to time, but unlike most hibernators, when it awakens it eats some of the seeds and berries it collected in the fall.

Stories in the snow

One day in early October the sky fills with leaden gray clouds, and a mass of cold arctic air moves in from the

When snow covers the ground, tracks in the fluffy surface reveal the comings and goings of animals that are rarely seen. Here the miniature squirrellike tracks of a deermouse are stitched together by the impressions made by its long tail.

northwest. At first sleet pelts noisily against the windows of the laboratory, and then snow begins to fall. It is a light snow—only an inch or so—but by the time it stops, the world is completely transformed.

In the forest, the ground between the trees is blanketed with white, although beneath the trees themselves it is still bare. Like umbrellas, the needle-covered branches of the pines have trapped and held the snow falling on them. Ecologists have adopted an Alaskan Eskimo name for this snow retained by the trees. They call it *qali*. The bare snow shadows beneath the trees, in turn, are called *qamaniqs*, while the snow on the ground is called *api*.

And what a fascinating story is written on the api! It is crisscrossed with the tracks of every animal that moved during the night. If you go for a walk in the forest now, you will find lots of evidence of the presence of taiga animals that remain active all winter.

Snowshoe hare tracks seem to be almost everywhere. Like other rabbits and hares, their large hind feet straddle the land well in front of their small forefeet when they jump. Similar tracks, but only about half the size, were made by red squirrels running from tree to tree. Another jumper that makes even smaller squirrellike tracks is the

Like dotted lines, a skein of tracks marks the wanderings of a pack of wolves across the snow-covered surface of a frozen lake. Even after the wolves have passed from sight, a skilled tracker could unravel the details of their daily prowling from these clues in the snow.

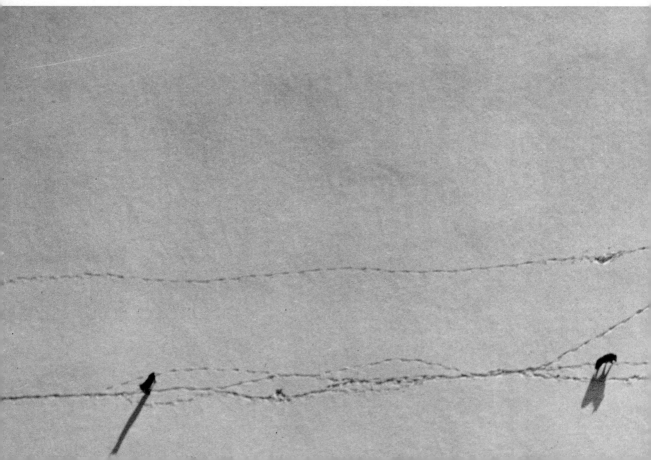

deermouse. However, its trail usually includes a long impression where its tail hit the snow.

Near some cranberries is a different kind of small trail. In some places it looks as if the maker simply plowed through the snow, but where it ran on the surface, it left two lines of evenly spaced footprints that look like the stitching on a seam. These are the tracks of a red-backed vole. Now that mushrooms have frozen, the voles have turned their attention to patches of persistent berries, especially cranberries, and this is where you can expect to find their tracks.

In one place you find a vole trail that ends abruptly with signs of a scuffle and a few drops of blood on the snow. Obviously the vole was killed by some larger animal whose tracks at first glance look as if they include the marks of only two feet. Look closely, however, and you will see signs of the other two feet. The predator was a short-tailed weasel or ermine. When it runs, its hind feet come down and land in almost exactly the same spots where its front feet were just a fraction of a second before, leaving what appears to be the marks of only two feet.

It might be a good idea to go out to the highway before too many cars and trucks pass, since many animals use

highways as easy travel routes on their nightly journeys. Today you are not disappointed. One shoulder of the road is trampled with what look like the prints of large dogs. During the night a family of wolves must have passed this way.

Where you first pick up their trail, the tracks are so intertwined that you cannot tell how many animals there were. Farther down the road, however, the wolves fanned out, apparently to investigate something that attracted their attention in the ditch. Now you can see that there were six wolves in all, probably a pair of adults and four surviving pups that were born the previous spring. It is impossible to tell from the tracks which were adults and which were young. As with young dogs, the feet of wolf pups are full size long before their bodies are fully grown.

Where the highway passes through a stand of black spruce, the tracks of still another animal cross the road. Though rounder, they are about the same size as wolf tracks. But unlike the wolves, the animal left no claw marks. So it must have been some kind of cat, for only members of the cat family can retract their claws into sheaths when they are traveling. The maker must have been the one cat of the north woods, the lynx. Evidently a

lynx spent part of the night hunting hares in the dense black spruce thicket.

If you go to Heart Lake, the small lake for which the biological station is named, you will find still other tracks. One end of the lake is marshy and overgrown with cattails, sedges, and bulrushes. Here and there the surface is dotted with dome-shaped muskrat homes, now capped with a layer of snow. In winter, the muskrats escape the cold by remaining in their snug houses and feeding on plants in the unfrozen water beneath the ice. Early in the season, however, they still venture out occasionally on top of the ice. Wherever they go, their trails are unmistakable. Between their two rows of footprints there is always a nearly continuous wavy line drawn in the snow by their long dragging scaly tails.

You should also be on the lookout for mink tracks. Mink prey heavily on muskrats and are likely to go from muskrat house to muskrat house in search of unwary victims. Mink are members of the weasel family, and, not surprisingly, their tracks are simply larger versions of ermine tracks.

Out on the lake is another set of tracks coming right across the open ice. In this case, the tracks are nearly the same size as a wolf's track; each foot has five toes, and

A pair of young wolves playing in the snow (*left*) leave evidence of their fun. Above, clawed wolf tracks are clearly etched on a snowy surface. The wolf's large front feet often leave tracks as much as five and one-half inches long.

The trail of a fisher wanders through a coniferous forest, its favorite habitat (above left).
The polar bear that left the tracks (above right) *was dragging its feet, with its claws leaving slashes in the snow between the footprints.*

PATTERNS IN THE SNOW

The snow is a perishable page, but before it melts or is covered by new snow, all who cross its surface print the record of their passage. With experience it is possible to read these stories in the snow, for although tracks come in all shapes, sizes, and patterns, those of each species are as distinctive as the fingerprints of an individual. To a novice, the maze of tracks and trails at first may seem as mysterious as hieroglyphics. But to a skillful tracker they are a code that reveals the story of the life that pulses in the far north even in the depths of winter.

116

Above, a red fox loped along in a characteristically straight and narrow path. Hindered by deep snow, a pack of wolves all plowed along in the same path, as domestic dogs often do (above right). To the right is the snowshoelike mark left where a flying squirrel landed after soaring down from the treetops. The three points at the rear were made by its hind feet and the base of its tail, while the separate mark was made by the tip of its tail.

there are toenail impressions. The tracks do not look the least bit weasellike—among other things, the hind feet do not step in the marks made by the front feet—but the maker in fact was the largest member of the weasel family, the wolverine.

You are lucky to have seen its tracks. You are not likely to see a wolverine in the flesh unless you spend a good deal of time in the woods, for wolverines are one of the rarest animals of the far north. And trappers are making them increasingly rare. For one thing, they often learn to follow a trapper's trail and remove or destroy everything caught in his traps. They also have a reputation for breaking into cabins and destroying food and furniture. Thus, trappers like to get rid of any wolverines in their vicinity. Their pelts, moreover, bring a good price. They are especially valued for trimming on the hoods of parkas since, unlike other furs, wolverine trim easily sheds the frost that forms when warm, moist breath comes in contact with cold fur.

If you could continue your walk, you might find the tracks of still other animals you are not likely to see in the flesh. Pine martens, for instance, live around Heart Lake, but they are secretive and active mainly at night. Where they walk through the snow, however, their minklike tracks provide an unmistakable record of their comings and goings.

Tracking, in fact, is an ideal way to learn of the activities of wild animals. Certainly your walk today offered convincing proof that even in winter the woods are alive with far more animals than you might have suspected were ever present.

Making your own heat

Within a day or so the first snow melts, and the blankets of white deposited by subsequent light snowfalls are just as ephemeral. Gradually the cold arctic air takes a firm grip on the land. Before long the temperature ceases to rise above freezing except at midday. Then, one morning in late October, you wake up to find a fresh two-inch blanket of snow on the forest floor. This snow will stay. Winter has arrived in the far north.

Now that temperatures are almost continuously below freezing, it is a good time to answer a few questions about

FRONT

HIND

The lynx's tracks look much like those of a very large house cat. Like all cats, it has retractable claws that never leave marks. Since its feet are quite heavily furred, its tracks in snow usually are indistinct.

118

exactly what the cold means to animals of the far north. The first thing to understand is that cold is simply the absence of heat. In a physical sense, we can only talk about the ways in which animals gain and lose heat.

One thing you are sure to notice is that birds and mammals are the only kinds of animals that remain active in cold weather. Like human beings they are *warm-blooded,* or able to maintain a constant body temperature of about 98 degrees despite fluctuations in external temperatures. *Cold-blooded* animals, in contrast, lack this ability, so that their temperature varies with that of their surroundings. Thus, when temperatures drop, all the insects, frogs, toads, snakes, and other cold-blooded animals living in northern regions must retreat to sheltered places where they remain in a state of dormancy throughout the winter.

Warm-blooded animals, of course, get the heat they need to keep themselves warm from the food they eat, but the process is a bit more complicated than it appears at first glance. Heat is a form of energy which obeys several precise physical laws. One rule is that energy cannot be created or destroyed; it can only be transformed from one kind of energy to another. The chemical energy stored in food material, for example, can be used to make muscles contract, thus producing the energy of motion.

Another general rule about energy is that no transformation from one kind of energy to another is ever one hundred percent efficient. Part of the energy is always "lost" in the form of heat. As long as an animal is alive, therefore, it is gaining heat as a by-product of the chemical reactions going on in every cell of its body. So, if one way of coping with the northern winter is to escape the cold by migrating or hibernating, another solution is to produce your own heat to keep warm.

Conserving heat

Manufacturing one's own heat calls for constant replenishment of energy sources, since a warm-blooded animal's body does not continue to warm up indefinitely. It is constantly losing heat as well as producing it. Just as soup cools when you pour it into a bowl and leave it on the table, heat constantly flows outward from an animal's body.

The wolverine is so rare and elusive that few people ever see more than its tracks. The five toeprints and distinctively shaped heel pads make its tracks easy to recognize, although sometimes the small fifth toe leaves no mark.

FRONT

HIND

119

This can happen in two ways. When heat energy is transferred from one molecule to another with which it is in contact, the process is called *conduction*. This is how the bowl gets hot when you pour soup into it—and why your fingers get warm if you touch the bowl.

However, heat is also being transferred by *radiation*. In this case, instead of passing from molecule to molecule, the heat is moving in the form of rays or waves. The most familiar example of radiation is the heat from the sun, which constantly passes through millions of miles of space in the form of energy waves.

Here then is a problem for an animal with a high body temperature living in a cold environment. By midwinter,

A lush coat of fur helps to keep the red fox warm in winter. Like beavers, martens, mink, and many other mammals of the far north, it is sought by trappers for the sake of its insulating pelt.

when the thermometer plummets to 60 degrees below zero, the difference between the animal's internal temperature and that of the surrounding air will be as much as 150 degrees. To cope with these conditions the animal must either step up the rate of life processes in its cells to gain more heat or slow down the rate at which it loses heat to the environment.

The best way to cut down on heat loss, of course, is to cover your body with a layer of *insulation* (a substance which is a poor conductor of heat) or a *radiation shield* (something that blocks the passage of radiating heat waves). Mammals have an insulating body covering in the form of fur, while birds are covered with feathers. Both are excel-

Like the fox's fur, the gray jay's feathers are superb insulators. The sleek tips of the feathers form a waterproof coat, while air spaces among their downy bases trap and retain the heat escaping from the bird's body.

lent insulators. In the case of mammals, their coats usually include a dense woolly layer of underhair covered by a layer of longer, coarser guard hairs. The sleek, oily guard hairs provide a protective waterproof cover, while the underhair traps heat in air spaces between the hairs. The down and outer feathers of birds act in exactly the same way in trapping and retaining body heat.

Caribou and other members of the deer family have a slightly different—but just as effective—kind of fur coat. Although they have little or no underhair, the individual hairs of their coats are filled with minute air spaces, and air is a poor conductor of heat. In addition, caribou hairs are actually thicker near their tips than at their bases. Thus the overlapping tips form an almost airtight shield, while the bases are surrounded by heat-trapping air spaces.

Another way to cut down on heat loss is to reduce body extensions which would radiate heat. Thin projecting ears filled with blood vessels, for example, are excellent radiators. That is why we wear earmuffs in winter. Fingers also radiate a great deal of heat, and so we wear mittens to keep them warm. Wild animals obviously cannot wear mittens, but many of them do have fur or feathers on their feet. The soles of snowshoe hares and arctic foxes are completely covered with fur in winter, and even polar bears' feet are almost completely furred. Even ptarmigans wear "socks" of feathers and have a feather covering on their toes.

It would also be helpful for northern mammals to have small ears or ears buried in their fur. And many of them in fact do have smaller ears than their southern relatives. Foxes illustrate this trend very neatly. The kit fox of southwestern deserts seems to be all ears—but in such a hot climate it is advantageous to be able to get rid of excess heat. Red foxes of moderate to cold climates in contrast have medium-sized ears. The arctic fox of the tundra, however, has such short, rounded ears that its face hardly seems foxlike at all.

The red-backed voles that we met earlier also are cold-adapted in this way. Their ears, buried in dense fur, are

Even in the bitterest cold, the arctic fox remains snug. Its tiny ears are insulated with heavy fur, as is the fluffy tail which it can wrap around its body like a comforter. Even its paws are covered, top and bottom, with fur.

Foxes illustrate the tendency of many kinds of animals to have smaller heat-radiating extremities in cold climates. In contrast to the small-eared arctic fox, the red fox, which lives in more temperate climates, has medium-sized ears, while the desert-dwelling kit fox has enormous ears.

ARCTIC FOX

RED FOX

KIT FOX

Stocky builds and luxuriant coats help keep muskoxen warm.

DEER MOUSE

RED-BACKED VOLE

The slender deer mouse's large ears and long tail radiate so much heat that in very cold weather it often spends days at a time in its nest in a torpor resembling hibernation. The red-backed vole, with its compact shape, short tail, and small ears, on the other hand, is better equipped to cope with the cold and it remains active throughout the winter.

all but hidden, and their tails are short. And, as winter progresses, we will find that they remain continually active. Deermice, on the other hand, slow down greatly in their activities and even slip into a deep sleep resembling hibernation for days at a time. Their prominent ears and long, thinly furred tails undoubtedly have something to do with this. Because of these large heat-radiating surfaces, keeping warm is a much greater problem for them than for voles of about the same size.

Strangely enough, even the size of an animal influences its ability to keep warm. The greater an animal's heat-losing surface area is in relation to its heat-producing bulk, or volume, the more difficulty it will have in coping with the cold. This relationship does not work at all as most people suspect, however. A mouse, with its small surface area, is actually at a disadvantage when compared with a moose, since, in proportion to its volume, the mouse has a much greater surface area.

This may seem hard to believe, but it is easy to convince yourself. Imagine a box that is a perfect cube one foot long on each side, and another that is two feet long on each side. The first box has a volume of one cubic foot and a surface area of six square feet. The larger box, however, has a volume of eight cubic feet (two feet by two feet by two feet) and a surface area of twenty-four square feet (four square feet on each of its six sides). The ratio of volume to surface area of the small box, then, is one to six, while that of the larger box is eight to twenty-four. Thus, if the boxes were animals, the larger box could contain eight times as many heat-producing muscles and glands as the smaller one. But its heat-losing surface would be only four times larger! Knowing this, it will be interesting to see what becomes of the smaller animals when it gets really cold.

The white shield

As autumn progresses, storms come in rapid succession. By early November the forests around Heart Lake are completely blanketed with snow. If you carefully dig a hole down to the moss on the ground, you will find that the snow is now ten inches deep. You will notice however that the texture of the snow on the top is different from that of the snow at the bottom.

126

With a small wooden stick, carefully remove a few snow crystals from the surface and examine them through a magnifying glass. Many of them will have lost an arm or two, and some will be starting to lose their shape, but they still will be recognizable as typical six-sided snowflakes. If you examine some crystals from halfway down the hole, however, you probably will not find anything even remotely resembling a snowflake—just small, irregularly shaped lumps. Crystals from the bottom of the snow are even larger icy lumps frozen to each other. Evidently the snow changes when it lies on the ground, and the longer it lies, the more it changes. Unlike the delicate fluffy flakes on the surface, the snow on the bottom has become a crumbly mass of icy pellets. There is even a trace of an air space between the moss and the snow.

If you place a thermometer at the lower surface of the api and leave it there for a few minutes, you will no doubt be surprised by your findings. Although the air temperature is 18 degrees below zero, under the snow the thermometer reads plus 25 degrees—just a little below the freezing point! Obviously the world beneath the snow is quite different from the one we live in. Today, for example, it is 43 degrees warmer under the snow. Moreover, if you were to check the temperature periodically over the course of twenty-four hours, you would find that it varies by no more than 1 or 2 degrees. Air temperature, in contrast, fluctuates by several degrees each day from its coldest point, usually in early morning, to its warmest reading, usually late in the afternoon. Besides being relatively warm, the *subnivean* (under the snow) world has a nearly constant temperature.

The reason for this is that the earth has a hot center. Even when it is frozen at the surface, it continues to radiate heat into space. Snow, however, is a good insulator because so much air is trapped between the interlocking arms of snowflakes. Thus, a thick blanket of api forms an excellent shield blocking the escape of the earth's heat.

This heat also accounts for the changing texture of snow as it lies on the ground. By the process of *sublimation*, water can pass directly from its solid state (ice) to its gaseous state (water vapor) and from vapor to ice without passing through the intermediate liquid state. As heat escapes from the earth, snow at the bottom of the api does not melt into water but instead loses vapor by sublimation. The vapor then passes up through the snow and refreezes

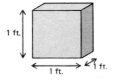

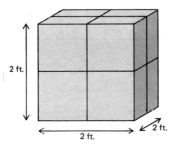

Large animals can conserve heat better than smaller ones because their heat-radiating surface areas are smaller in proportion to their volumes. A cube one foot long on each side for example has a volume of one cubic foot and a surface area of six square feet, while a cube two feet long on each side has a volume of eight cubic feet and a surface area of twenty-four square feet. Thus the larger cube has eight times as much volume as the smaller one, but only four times as much surface area.

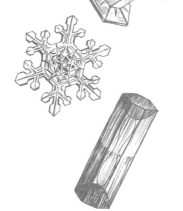

on crystals at intermediate layers. The constant upward movement of sublimated vapor thus not only accounts for the granular quality of the deeper layers of snow, but, more importantly, results in the formation of an air space between the snow and the ground.

What all this adds up to is an explanation of how small animals such as voles and shrews are able to survive the northern winter. Instead of escaping the cold by migrating or hibernating, they avoid the worst of the weather by living under the thick insulating blanket of api. Despite its small size, a vole can fare as well as a moose because it rarely exposes itself to the temperatures a moose must endure. The air space beneath the snow, moreover, allows it to move about freely in its search for food.

Voles, of course, are not the only ones to benefit from the snow. Ermines and, to some extent, martens also spend a lot of time hunting under the snow in late winter. In addition, the snow protects dormant insects, spiders, snails, and even the roots of plants from the most severe effects of the far-northern winter. Even ruffed grouse take advantage of the insulating qualities of snow. Rather than roost out in the open, they bury themselves in the snow at night. The birds fly down, and as they land, dive into the soft snow at an angle, then tunnel in for a foot or two. There they spend the night in comparative warmth and, when morning comes, burst straight up from their snow caves and fly off to feed.

A cold Christmas

By Christmas, Heart Lake is really in the icy grip of winter. Eighteen inches of snow cover the ground, and it is bitterly cold. Temperatures near 40 degrees below zero are common now, for the whole area lies under an immense dome of cold, dense arctic air. The sky, moreover, is clear, with no clouds to act as radiation shields blocking the escape of heat from the earth. So, despite the blanket of snow, the earth each night radiates more heat to outer space, while the weak sun by day cannot make up for the loss. The sun now is up for only four or five hours a day, and even at noon barely rises higher than the treetops on the horizon.

Fortunately there is no wind, for that would make the cold even more intense. Besides blowing away the layer of warm air trapped in your clothes, wind would step

By midwinter, the far north is completely blanketed with snow. The snow on the ground is called api, while the clear spaces and hollows beneath trees are termed qamaniqs. Snow caught on the branches, in turn, is referred to as qali.

up the evaporation of any moisture on your body, cooling it even more. It is probably to escape from wind that many small winter birds of the north woods take refuge in tree cavities. Just as grouse tunnel into the snow to keep warm, chickadees, nuthatches, and woodpeckers all resort to nest holes in trees when they are not out feeding. In really cold weather, moose also head for sheltered ravines or go into dense stands of spruce where air movement is less. Even snowshoe hares tend to hide in the tentlike spaces beneath qali-laden spruces now. Besides cutting down on wind, the snow-covered branches form a radiation shield which slows down the heat loss from their bodies.

Curiously enough, animals now are also faced with the danger of becoming overheated. Although it is so cold that trees are cracking with reports that sound like the shots of a small rifle, you will have to move at a slow steady pace if you go for a hike today. If you work too hard, you will begin to perspire inside your warm winter clothes. Later, you will become chilled when body heat is used up as the perspiration evaporates. Some of the moisture, moreover, will freeze when you relax and so reduce the insulating power of your garments.

Wild animals also produce more heat than they need to keep warm when they work hard in winter. When wolves chase caribou, for instance, they produce excess amounts of heat, as do sled dogs when they are pulling sleds. How do they avoid becoming overheated? Sled dogs have been studied in this respect and have provided a partial answer.

In winter, the willow ptarmigan grows fluffy stockings of feathers on its legs and feet. Besides providing insulation, the feathers transform its feet into tiny snowshoes, helping the bird pad along on the surface of the snow without sinking in.

For one thing, they have few or no sweat glands on their bodies to wet their hair. Furthermore, they can flatten their hair against their bodies and so reduce the thickness—and efficiency—of the insulating layer. In addition the insulating value of the fur on the dog's leg is much less when the leg is moving than when it is still. Bending the joints apparently causes the fur to spread apart and lets bare skin come in contact with the surrounding air. Finally, dogs pant with their mouths open and tongues hanging out. Cold air blowing on the tongue carries away some excess warmth and more is lost as moisture evaporates from the surfaces of the mouth and tongue.

Of snowshoes and stilts

Although it is a fine day for a hike, with eighteen inches of snow on the ground you will have a hard time getting around unless you wear snowshoes. By distributing your weight over a large area, snowshoes enable you to walk without sinking more than a couple of inches into the fluffy snow. A good many animals of the far north also solve the problem of getting around in deep snow by growing their own snowshoes. The most obvious snowshoers are snowshoe hares, and, on the tundra, arctic hares. Their enormous, well-furred hind feet are such efficient snowshoes that they seem practically able to float over the surface of the api. In addition, their toes spread apart when they run, increasing the

The snowshoe hare not only turns white in winter but also grows insulating snowshoes of fur on its large hind feet. Like the hare's snowshoes, the skis and snowshoes worn by humans function by distributing the wearer's weight over a larger area of the snow's surface.

SLED DOGS: POWERHOUSES OF THE NORTH

For centuries, sled dogs have been man's traditional means of winter transport in the snowy arctic. Nowadays airplanes, snowmobiles, and all-weather vehicles are making inroads into the dogs' dominance, but the stout huskies still remain the best-adapted natural means of transportation. They are able to travel when planes are grounded, and they never freeze up, run out of gas, or become brittle from the cold. Nor are the dogs laggards when it comes to work; given the right driver and the right wind and snow conditions, a team of twelve is said to be able to haul a load of a thousand pounds at a rate of three miles an hour.

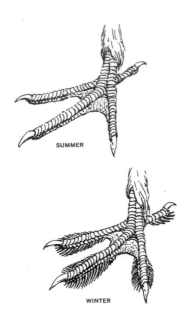

SUMMER

WINTER

While the willow ptarmigan grows feathery snowshoes in winter, the ruffed grouse grows comblike fringes of horny material on the sides of its toes. The fringes may help the grouse walk on snow without sinking in, but probably are more important in providing a strong grip on ice-covered branches.

snowshoe effect. Squirrels, in contrast, have more trouble. In soft snow you can hardly recognize their tracks: each place where a squirrel lands looks like the imprint of a single foot of a larger animal, since the squirrel's whole body sinks into the snow.

The hare's main predator, the lynx, is at no disadvantage in the snow, for it too wears snowshoes. Compared to other cats, its feet seem disproportionately large and, in winter, are completely covered, top and bottom, with fur. So, when the chase is on, the lynx, like the hare, is able to float over the surface of the snow.

Even caribou have a slight snowshoe effect. Besides being large to begin with, the two hooves on each of their feet tend to spread apart when walking, thus increasing the surface area.

If you hike among some willows, you are likely to discover some entirely different tracks, with three toes pointing forward and one pointing back. As you enter the willow grove, there is an explosion of sound as a dozen or so grouse-like birds take to the air and settle down again a little farther off. You did not notice the birds at first because, except for their black bills and tails, they are pure white. And their toes are completely covered with feathers, transforming their feet into snowshoes. The birds are willow ptarmigan. Although they nest on the tundra, many of them migrate, sometimes several hundred miles, into the taiga to spend the winter. There, camouflaged in their white plumage and able to walk on the snow as well as fly, they live in comparative safety among the willows, feeding on willow buds.

Ruffed grouse also have a kind of snowshoe, but instead of feathers they grow combs of horny material along both sides of each toe. However, there is some question as to whether they really serve as snowshoes. Some zoologists think these structures may be more important in helping the grouse to grip frozen and perhaps snow-covered branches of the trees and tall shrubs in which they feed.

In contrast to the snowshoers who float over the surface of the api, some animals of the far north wade through snow on stiltlike legs. Moose tracks often are simply holes punched through the snow. Their long legs work almost straight up and down, and the feet are lifted so high at each step that often the hooves do not even drag across the surface of the snow. Wood bison and caribou, on the other hand, tend to play follow-the-leader. One animal plows through the snow, breaking a path, and the rest plod along behind in its tracks.

The final way to get around in the snow, of course, is to tunnel through it. Voles burrow through the snow as well as in the air space beneath it. Here and there on the surface you will find neat round holes less than an inch in diameter leading to their sloping tunnels. You may even see tracks where a vole made a brief excursion into the cold outer world and then returned to subnivean darkness. Or perhaps you will see tracks that end at a flurry of wing marks on the snow where a great horned owl swept down to snatch the unwary wanderer.

Tunnels apparently are necessary for the voles, but no one knows exactly why. Some people have suggested that the tunnels act as chimneys, permitting the escape of carbon dioxide from the air space beneath the api. Carbon dioxide is the waste gas that all animals expel with every breath. Since carbon dioxide is poisonous in large amounts, it is important to get rid of it. Unfortunately for this theory, however, in the few instances in which scientists have carefully withdrawn air from under the snow, they have not found any large concentrations of carbon dioxide.

A second theory suggests that the voles come to the surface in order to get some cue as to the time from the sun,

Although they pop to the surface occasionally, lemmings, like voles, spend most of the winter in the relatively warm air space between the soil and the snow or in tunnels running through the snow itself.

In autumn, agile boreal chickadees tuck seeds and other bits of food among the lichens growing on the undersides of pine and spruce branches. Later, when snow covers other food supplies and qali covers the upper surfaces of the branches, the chickadees return to retrieve food they stored in the fall.

moon, or stars. Although no one has proved it, this seems like a possibility. Under the snow, the voles live in total darkness. Yet studies have shown that they are most active just after sunset. How do they know the sun has set unless they get some kind of cue from the outer world?

Winter food stores

As you continue your hike, you are sure to notice many signs of feeding activities. To produce the large amounts of heat they need to keep warm, animals that remain active through the winter must find nourishing food to eat every day. The snow, however, has covered a lot of food that was readily available in summer, forcing many animals to make do with whatever they can find.

As you watch, for example, a chickadee flits into a nearby pine and begins to work its way, upside down, along the underside of a main branch. Flicking its wings and fanning and folding its tail for balance, it probes constantly with its bill under flakes of lichen or between the needles. The bird's jerky movements dislodge puffs of snow from the branch, but it seems hardly to notice the inconvenience as it continues its search for food.

It is lucky for the chickadee that it is so agile and can hunt upside down when the upper surface of the branch is covered with qali. Or is it luck? If you think back, you will remember that chickadees around the lab were acting like acrobats ever since your arrival in the fall—even when the tops of the branches were not covered with snow. Could it be that the chickadees were storing food on the undersides of branches, as if "knowing" they would soon be reduced to foraging there?

If you take a branch back to the lab and examine it closely, you will find that this is exactly what the birds were doing in the fall. Hidden here and there are all sorts of food items that could not possibly have lodged on the branch by accident. Most of the food is stored on the twigless, lichen encrusted portion of the branch near the main trunk. Tucked among the flakes of lichen are winged seeds of spruce and pine, seeds from the fruits of junipers, and even a few small insect pupae. Some food is also stored near the tip of the branch, where the jack pine needles grow in pairs.

136

A favorite trick of the chickadee is to wedge pine seeds between the bases of the paired needles.

Although chickadees do not depend entirely on stored food, obviously this is a good way to ease the problem of finding enough to eat in winter, especially when days are so short that all the food for a day must be found in just a few hours. In the Old World, willow tits, as chickadees are called there, go a step further. One Norwegian study has shown that willow tits store enough food to last them throughout the snowy winter months.

Red squirrels also store food while it is still plentiful in autumn. The squirrel we saw on our first hike at Heart Lake, for instance, was caching pine cones about its midden. Although we will not hear any of its calls or find any fresh tracks if we return to the midden today, the squirrels are still active. If you dig carefully into the snow on top of the midden, you are likely to find a few tunnels about the right size for a red squirrel. Like the voles, they have finally taken refuge under the api, for the rules of heat loss and body size are now operating: no mammal smaller than a snowshoe hare is any longer able to endure the temperatures above the api. But even if it is too cold for the squirrels to go up into the trees for cones, they are still able to survive on the ones they stored in the fall.

Red-backed voles find plenty of food as they tunnel through the air space under the snow. In addition to feeding on seeds, grass stems, lichens, and other food on the ground, they gnaw on the bark of shrubs that they encounter on their travels.

Food underfoot, food overhead

Although they do not store food, voles have no particular problem finding food. As they travel about in the air space beneath the api, they eat grass stems, gnaw the bark from shrubs, and eat any other bits of plant food they find along the way. When they tunnel up through the api, they also find tree seeds that blew on the surface and were buried by later snowfalls. Even lichens and other material that rain down when a lump of qali breaks loose and falls from a branch form welcome additions to their diets.

But the mat of vegetation beneath the api is unavailable to most animals. Only a few dig through the snow to get at it. The most obvious are caribou. With their large hooves, which are nearly five inches across, they paw the snow aside, forming dishpan-sized basins. Then they eat the lichens they have exposed on the ground and move on to another spot. In winter they also vary their diets by feeding on the beardlike lichens that hang from branches in the boreal woodlands. Wood bison also continue to eat vegetation on the ground. Instead of pawing the snow aside with their feet, however, they swing their heads from side to side and clear away the snow with their muzzles.

Up on the tundra, the muskoxen use quite a different approach. They let the wind do most of the work of digging out their food. In summer, muskoxen live among the groves of willows in gullies near streams and rivers, feeding on willows, glandular birches, and other plants. In winter, when deep snow drifts into the sheltered low spots, it buries most of the vegetation and makes it difficult for the short-legged muskoxen to walk. So when winter comes, the herds move to higher ground where wind has blown away much of the snow. In some places, the wind has cleared away all the snow, exposing the mats of low creeping tundra plants. Where some snow still remains, it is shallow enough so that the muskoxen can paw it away to get at their food.

Reindeer, like caribou, paw basinlike holes in the snow to uncover lichens and other vegetation. Their large feet also serve as snowshoes, since the two halves of the hooves spread apart when they walk. In addition, when the large hooves break through a crust on the snow, they make holes big enough to protect the animals' legs from abrasion as they walk.

Other planteaters also continue to get food wherever they can find it—which is mainly off shrubs and saplings that project beyond the api. Moose, for example, supplement their main diet of tree leaves and twigs with grasses and water plants in summer, but in winter they must live almost exclusively on willow, birch, and aspen twigs. With their long legs, they have no problem reaching the branches. However, they have one feeding method that makes high branches available to smaller animals as well.

If you follow a moose trail long enough, you are likely to come eventually to a broken sapling. In order to get at tender twigs out of its reach near the top of the tree, the moose pushed against the sapling with its chest. As the sapling bent, the moose walked over it, nibbled on the twigs, and then moved on. In summer the supple young tree would have sprung upright as soon as the moose left. In winter, however, the trunk was brittle with frost and snapped when the moose pushed it over. Now the treetop lying on the ground provides a banquet for snowshoe hares, which nibble on the twigs and bark, and for ptarmigans who eat the overwintering buds and remaining seeds.

When it comes to finding food in winter, the wood bison really uses its head. Swinging its big muzzle from side to side, it sweeps away the snow to get at the dried grasses and other plants buried beneath the surface.

Qali also brings treetops within their reach. Just as the evergreens droop beneath the heavy loads of snow on their branches, the dense twigs on young deciduous trees catch a great deal of qali. Sometimes the tops are bent so low that they nearly touch the api. Here, too, hares and ptarmigans take advantage of this new food source. When the snow melts and the trees spring upright again, they look odd indeed with bark gnawed from the tips of the highest branches, well beyond the reach of the tallest animal in the forest. Alaskans, noting the appearance of the trees in spring and knowing the cause, call snowshoe hares "high bush moose!"

Hares and ptarmigans also benefit from the presence of the snow itself. They can easily walk about on the surface without sinking in. Thus as more and more snow accumulates, they are continually being brought within reach of new food supplies in the form of bark, twigs, buds, and seeds at higher and higher levels above the snow.

The hunters

As for predators, their lives continue pretty much as usual in winter. The snow may cause them to work harder for a meal now, but as long as their prey species remain healthy and plentiful, they continue to hunt much as they do in summer. Wolves still follow on the trails of moose and caribou, and lynx course over the snow in pursuit of hares. Martens run down red squirrels in the treetops, and also hunt for hares, grouse, and ptarmigans.

Life is a bit more difficult for predators dependent on small animals that live under the snow. Besides being warm, the world beneath the api is comparatively safe. Unable to dig beneath the snow, owls must now be constantly on the lookout for hapless voles that venture out on the surface of the snow. Otherwise, they will have to turn to other foods. Great horned owls, for instance, sometimes kill hares, while boreal owls often must be content with small birds.

By May the willow ptarmigan is partially covered with its variegated summer plumage. Throughout the winter, the accumulating snow has continually brought it within reach of the buds of willows that were far above its head in autumn when the ground was bare.

The lynx is a skillful hunter of hares and other small prey, but occasionally accepts the role of a freeloader if it encounters the carcass of a deer or caribou left unfinished by wolves or other predators.

140

Red foxes, on the other hand, do manage to get at voles as long as the snow is not too deep. Apparently they can detect the presence of voles in shallow snow, perhaps by scent or possibly by hearing the tumbling of coarse snow crystals dislodged from the bottom of the api by the voles' movements. In any case, once a fox detects a vole under the snow and calculates its position carefully, it makes a high leap and comes down on all four feet on just the right spot. If the fox's radar has been accurate, the vole will be pinned underfoot ready to be eaten. Or, if necessary, it will dig through the snow to get at voles.

For ermines and least weasels the hunt is even simpler. Their lithe bodies are so slender that they can pursue voles beneath the snow in their own tunnels.

Winter white coats

Ermines, of course, turn white in winter when their world turns white, as do many other northern animals, which raises an interesting question. Do their white coats serve as camouflage, making predators less conspicuous as they stalk their prey and making the prey species more difficult to detect? Since white coats are such a widespread phenomenon in the far north, this certainly seems like a convincing possibility.

Willow ptarmigan, rock ptarmigan, and white-tailed ptarmigan, for instance, all wear mottled brownish plumage in summer, colors that blend effectively with the tundra vegetation. In winter, the rock ptarmigan is pure white except for its black tail feathers and black marks across its eyes. The willow ptarmigan has black feathers only on its tail, while the white-tailed ptarmigan wears no black at all. Thus, when snow covers the ground, the birds again are practically invisible. All in all, they seem to make a convincing case for protective coloration—except for one thing. The closely related spruce grouse and ruffed grouse also winter in the taiga, yet they wear their dark plumages year round. And they survive, possibly because they spend much of their time up in trees rather than on the snow-covered ground as ptarmigans do.

The same is true of lemmings. Brown lemmings remain brown winter and summer, while collared lemmings molt to white coats in winter. Yet the brown lemmings seem to be at no disadvantage.

The situation with predators is just as inconclusive. Ermines and least weasels are dark in summer, white in winter. Yet no other far-northern members of the weasel family—mink, martens, fishers, and wolverines—turn white in winter. The situation with arctic foxes is equally confusing. Most are tawny in summer and white in winter, yet in some places some of them—the so-called "blue foxes"—turn

In summer, the short-tailed weasel's chestnut back and yellow-tinged belly blend well with its brushy and rocky surroundings (*left*). Like many animals of the north, however, it sheds its summer coat for a nearly invisible white one when snow begins to fall (*right*). In its winter coat, it is known as ermine and is trapped for its regal fur.

Gyrfalcons, handsome circumpolar birds of prey, often are nearly pure white flecked with black. But many of them are almost completely black, while others are feathered in various shades of gray. Regardless of color, however, all are equally successful as they hunt for small mammals and birds.

a grayish-blue color in winter. Wolves also may be gray, brown, black, or cream-colored, but in some places are pure white. Only the polar bear remains indisputably white or off-white in all seasons and in all places.

Nor are white animals restricted to arctic regions. Snowy white herons and egrets thrive in the tropics, and white gulls and terns live in all sorts of aquatic environments. Yet the preponderance of winter white animals in the far north must have some significance. Surely the white feathers of snowy owls and gyrfalcons and the white fur of snowshoe hares, arctic hares, and many other creatures have some survival value, but it would seem that there is no easy explanation as to what it may be.

A break in the weather

Although days begin to lengthen in late December, the increase at first is imperceptible. By early February, however, the sun sets noticeably later every afternoon. The cold

144

weather also begins to ease a little. During cold spells in January, the thermometer slid to minus 50 once, and even the temperature under the snow dipped to 10 degrees. During the extreme cold, there were few fresh tracks on the snow, for animals were moving about as little as possible. Ever since, however, temperatures have been inching slowly upward, and the forest has become alive again with increased activity.

Because there has not been a wind since the country fell under the grip of the arctic air, the trees are now heavily laden with qali. But today a slight wind is moving in from the distant Pacific Ocean; the air is softening; and clouds are drifting in from the southwest.

As the day progresses, rising temperatures and freshening wind gradually do their work. The grip of the qali on the needles of spruces and pines is slowly weakened, and soon the snow begins to fall. First one branch is relieved of its load. Then as it springs upwards, it bumps other branches and dislodges their burdens. Falling lumps of qali from upper branches trigger minor avalanches on lower branches, and soon snow is plummeting from an entire tree, carrying with it needles, twigs, seeds, lichens, and other food for animals on the ground. When it is over, a light plume of powder snow drifts downwind from the newly liberated tree.

One after another the trees shed their burdens. The air is laden with tiny ice crystals, and the forest is full of the sound of dense lumps of qali plunging into the api and cratering it like the surface of the moon. It is a welcome sight to behold. Temperatures are still well below zero. But the worst of winter is now over.

Winter birds

With the change in weather, birch trees have shed the seeds from between the scales of conelike structures that dangle from their branches. Beneath each tree the snow is dotted with the tiny brown specks. Rising temperature and increasing humidity apparently caused the catkins to open and release their precious contents.

Though tiny, birch seeds are a rich source of energy and so are eagerly sought by small birds. Chickadees, for one, come to the feast beneath the birches. Although they have

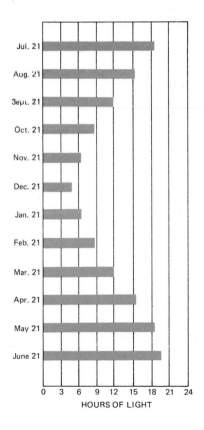

Comparison of the hours of daylight on the twenty-first of each month in central Canada shows that the short days of winter begin to lengthen rapidly as the season wanes. The shortest days, in December, are only five hours long but just two months later, on February 21, the sun remains above the horizon for almost nine hours a day.

145

their food stores, they cannot pass up a bonanza like this. Flocks of redpolls also swarm in to share in this newly available source of food.

If you stop to think about it, however, the presence of these sprightly active birds raises an interesting question. They are about the same size and weight as red-backed voles in winter. How is it that these small birds can survive above the snow at temperatures that would soon be fatal to voles? How can they keep warm enough despite their small size?

For one thing, their feathers provide extremely effective insulation. In addition, their food has a higher energy content than the voles'. By this time in winter, about two-thirds of the voles' food intake consists of lichens, and the rest is persistent fruits of cranberries and bearberries. Redpolls, in contrast, concentrate on birch and alder seeds. While lichens are relatively low in food value, seeds contain highly concentrated nutritious food stores laid down by the plant that will power rapid growth when the seeds germinate.

Ease of digestion also is important. Unless food can be digested quickly and carried throughout the body, cells will be unable to keep up with the demands made upon them to produce more heat. Here again, seeds have the edge over lichens. In addition, the redpoll's digestive system increases in efficiency in winter, a trait that has yet to be demonstrated in voles.

All in all, the redpolls fare quite well in spite of their small size. As long as seeds are plentiful, they can produce heat fast enough to maintain their internal temperatures. Food must be abundant, however, since the birds must take in enough in four or five hours of daylight to last through the following nineteen or twenty hours of darkness. Here again a special adaptation comes to their aid. Redpolls have a sack, something like the crop of a chicken, which they stuff with extra food just before darkness falls.

From time to time you are also likely to see robin-sized pine grosbeaks, especially if you stock the feeding tray near the laboratory with seeds. Their stout bills are admirably

Dark, persistent, conelike female catkins line alder twigs throughout the year. As with birches, the catkins contain minute but nutritious seeds that provide a source of winter food for many songbirds.

In spring, alders are tasseled with graceful inch-long catkins of tiny male flowers, which produce the pollen that fertilizes flowers growing in shorter conelike clusters elsewhere on the same shrub.

147

The twisted tips of the crossbill's beak are unique tools for prying open the cones of pines and spruces. The bird inserts its bill between the cone scales, then opens it, thus spreading the scales apart and exposing the seeds, which the bird scoops out with its tongue.

suited for cracking seeds, and they will eagerly consume your offerings. They are not a common sight in winter, however. Some remain in the northern taiga, but many more migrate to southern Canada and even to the northern United States.

A few white-winged crossbills may also show up from time to time. You are most likely to see them high in the spruces, busily prying open cones with their odd-looking crossed bills. Like the grosbeaks, most of them migrate south with the onset of winter, especially in years when the spruces fail to produce an abundant crop of cones.

Winter on the wane

With the arrival of March, the sun begins to regain its power, rising earlier and setting later day by day, and each day climbing higher into the sky at noon. The api still reflects most of the sunlight falling on it, but dark objects, such as bare trees, absorb more and more of the sun's heat energy. Cones, twigs, and other debris on the snow are warmed by the absorption of heat; as their increasing warmth melts the snow around them, they gradually sink into little depressions.

By late March the api reaches its greatest depth—thirty to thirty-six inches. Any additions of new snow now will be offset by settling of the old. Best of all, cold spells are becoming less frequent and less severe. Midday temperatures regularly creep up to zero and occasionally even reach the plus side of the thermometer.

As a result your daily hikes become increasingly pleasant. When the temperature reaches zero, you now feel positively warm. In addition the forest shows signs of becoming alive. Once again red squirrel tracks are to be seen everywhere, and the silence is broken now and then by their chatter.

Gray jays, which were seldom seen during the deep cold of midwinter, also become conspicuous again on sunny March days. By the end of the month, in fact, they will be nesting, though they are so shy that you are not likely to discover a nest. Why they nest so early is another unsolved northern mystery. If they produced a second brood of young later in the season, getting an early start would be an advantage. But they do not.

One thing that changes noticeably now is the surface of the snow. The warmth of the sun during the day followed by cooling at night causes a crust to form on the api. At Heart Lake the crust seldom becomes very thick or tough, but it does cause additional hardship for some of the animals that wade through the snow. Foxes and wolves, for instance, can now travel only by bounding. With each leap they break through the light crust and sink into the snow. Since this method of travel is difficult and tiring, you rarely see their tracks in the depths of the forest. Instead they prefer to travel on lakes, where even the moderate winds of the taiga are strong enough to compact the snow into drifts firm enough to support their weight.

The thaw

In April, a definite promise of spring is in the air. The sun now shines for more than twelve hours each day and really begins its work of destroying the api. Through the process of sublimation, molecules are being driven from the snow crystals directly into the air in the form of water vapor. At the same time, some of the surface crystals are melting into water that trickles downward through the api. When the water reaches a layer of snow that is still below freezing, it changes again to ice. In this way ice crystals in the middle layers of snow are cemented together into still larger granules. At first none of the meltwater soaks through to the mosses or the surface of the soil, since the temperature beneath the api is still below freezing.

So the animals that live under the snow are still safe in their subnivean world. However, their environment does change in one significant way. Light penetrates the old granular snow more easily than new fallen snow. The increase in light possibly is a signal to these animals that winter is almost over. In any case, voles begin to gain weight, and their reproductive organs begin to enlarge. They also become more active. Even deermice respond to the approach of spring. In contrast to their near dormancy during the coldest months, they now begin to wander about very actively each night.

Finally, one night in late April, the temperature never falls below freezing, and the snow continues to melt through-

The cylindrical hollow around the trunks of two trees is a sure sign that spring is on the way. The hollow formed because the tree trunks, instead of reflecting the sun, absorbed enough warmth to melt the adjacent snow. Fallen leaves and bits of bark gradually sink into the snow because they too absorb enough heat to melt a pit in the surface.

out the night. Meltwater from the surface does not refreeze as it soaks through the snow but instead seeps all the way down to ground level and wets the mosses and lichens. Now the spring thaw is on in earnest. After a day or two of this mild weather, pools of meltwater accumulate in low spots, flooding the nests and burrows of voles, mice, and shrews. As a result, some of the small animals are forced to abandon the places where they lived all winter and try to make a go of it on higher, drier ground. In some years, spring continues to advance with a rush of warm weather, and the crisis of the thaw is over in just a few days. In those years, the forest is once again free of snow by the time May is a few days old.

In other years, however, winter reasserts itself after the beginning of the first thaw. Temperatures plummet, and the pools of water freeze. An icy shield covers broad expanses

The spring thaw creates strange forms on the northern landscape. Here, great chunks of ice that were stranded beside the Thelon River are diminished drop by drop as the warming sun climbs higher in the sky each day.

of vegetation, locking it away from hungry animals. Scattered patches of old snow persist here and there, and may even be supplemented by new storms until the middle of May. But these poor remnants of the thick winter api have little insulating value and offer poor protection to animals of the forest floor.

At such times, voles and mice stop growing and suspend their preparations for breeding until winter has had its last fling. Their first priority is to try to stay alive, a challenge that consumes all the food energy they can obtain from their hostile environment. They have no surplus energy left over for such secondary objectives as growth or reproduction. In spite of their best efforts, many voles and mice perish in such years, either as a direct result of harsh conditions or because their desperate need for food exposes them more often to the watchful eyes of predators.

The thaw is a critical period for lemmings, voles, and other small ground dwellers. Should the meltwater freeze during a cold spell, ice would cover their food supplies and shelter would be hard to find.

Signs of spring

Each day you find that your daily walk in the woods becomes more difficult. If there is frost at night, it is easy enough to walk on your old, well-packed trails in the early morning before the daily thaw begins. But as the day progresses, bridges between the ice crystals disappear, and the snow crumbles underfoot without providing any support. Snowshoes now are virtually useless, yet it is still difficult to get around without them.

Squirrels do not seem to like these spring snow conditions any better than you do. The squirrels, however, are more active than ever, though most of their activity takes place in the trees. On your daily travels, you regularly see one squirrel noisily chasing another through the treetops. They are not fighting. Their noisy escapades are actually part of their courtship ritual. The red squirrels, like many creatures of the far north, are ready to breed.

As plants long buried beneath the snow begin to re-emerge, living also becomes easier for small birds. Continued melting causes qamaniqs to reappear beneath the trees, providing additional feeding areas. Soon the rigors of winter in the far north will be over, and animals of the taiga will again be faced with the comparative ease of surviving in the summer months.

Finally, one day in early May, the sky is filled with noise, and out of the south a ragged "V" of migrating snow geese appears, heading for the summer breeding grounds far north on the tundra. Now you know for certain that the long taiga winter is over. And now that you have seen something of how animals survive the brutal living conditions of the northern winter, you are likely to be tempted to follow the geese and find out what goes on in the brief summer of the farthest north.

Just as the arrival of robins is hailed as a sign of spring in more temperate areas, the first migrating Canada geese are greeted in the northern world as a signal that winter at last is coming to an end.

Land of

the Long Day

Snow still covers the ground when the snow geese arrive on their breeding grounds along the arctic coasts of Canada. As they wing in from the south, their shimmering white bodies contrast vividly with the clear blue sky. Then, like a flurry of giant snowflakes, they settle to the ground and, except for their black wing tips, become all but invisible against the blanket of white.

A good place to observe their arrival—and the coming of summer on the tundra—is at the mouth of the Anderson River where it flows into the Beaufort Sea on the Canadian coast about three hundred miles east of Alaska. If you arrive in mid-May, when the geese arrive, you will feel as if you are returning to midwinter. The temperature is still below freezing, and the surface of the api is sculptured into wind-blown drifts. Unlike snow in the taiga, the api here is so tightly packed that you can walk on top of it. Even the airplane that brought you to this remote northern outpost was able to land on the snow without sinking in very far.

However, there is one sure sign that winter cannot last

Flying in over ice and snow, black brants return in early June to their breeding areas on arctic islands and the northern coasts of Canada and Alaska. Later on, the small black geese will lay four or five buffy eggs in shallow, down-lined nests.

much longer. On your first night on the Anderson River, the sun dips only briefly beneath the northern horizon, then rises again. The long arctic day is about to begin. For the next seventy-five days until late July, the sun will remain constantly above the horizon.

Early birds

Snow geese are not the only birds you will find in this frozen landscape. At its mouth, the Anderson River fans out into a delta five miles across, a patchwork of low islands separated by winding channels of water. In May the entire delta is still blanketed with api, so that the islands appear only as white mounds on a field of white. As you scan the islands with your binoculars, you see a few pairs of large white birds with long graceful necks. They are whistling swans, already here to await the thaw. The smaller, darker

156

birds with shorter necks on some of the islands are white-fronted geese, also early migrants to the far north.

None of the birds seem to be doing very much. They just stand there on the snow. They are not feeding, because their food is still buried beneath the api. For the time being they must get their energy from fat stored up during the earlier stages of their leisurely northward migrations.

If you go upstream a mile or so where there is higher ground, you will find other birds that have been here all winter. Walking on the uplands will be more difficult. Where the tops of willows protrude from the api, the snow is so soft that you sink in up to your knees with each step. But if you stick to the hard-packed api between the willow clumps, you can move along quickly and easily.

Before long you find the familiar tracks of the birds you are looking for—willow ptarmigans. Suddenly, almost in front of you, about a dozen of them explode into flight, uttering their guttural "chur-ur-ur-ur." The ones with conspicuous

157

chestnut-colored heads and necks are males molting into their summer plumage. The females, however, are still almost completely white and nearly invisible against the snow and willows.

If you look at the few short willow twigs that project above the snow, you will find that they have been completely stripped of buds. Even twigs are scarce, for they too have been eaten by the ptarmigan. However, the willow flats are extensive, so the birds have had plenty of food all winter. Even the mosaic of snow types in this area is ideal for them. They can walk about on the hard-packed tundra api and can easily dive into the soft snow around the willows when they are ready to roost.

Perched on a windswept ridge nearby is another bird that winters here, a snowy owl. As you watch, the owl takes off and flies with slow, steady, silent wingbeats toward the place where the ptarmigan disappeared. Suddenly the owl plunges to the ground, and ptarmigans scatter in all directions. But the hunt has been successful. Through your binoculars you can see the owl beginning to tear feathers from the neck of a bird clutched between its talons.

In early spring, much of the tundra remains blanketed with deep snow. The branches of willows bristling through its surface (*left*), however, continue to serve as the staff of life for many arctic animals. Willow ptarmigan (*right*) pick off the buds, muskoxen and caribou eat the twigs, and lemmings and arctic hares nibble the bark and twigs. Even Eskimos collect the leaves and buds of certain species for a food rich in vitamin C.

By going up to where the owl was perched, you can find out what else it lived on during the winter. Owls eat their prey whole or in large chunks, swallowing bones, fur, and feathers along with the meat. Later, as they sit on their habitual perches, they regurgitate the indigestible material in large moist pellets. With experience, biologists are able to identify many of the remains and thus figure out what owls eat.

In this case, it comes as no surprise to discover that a few of the pellets are composed of white ptarmigan feathers. Most of them, however, are made up of the fur, bones, feet, and skulls of mouselike animals, for small rodents are the mainstay of the snowy owl's diet. Several of the pellets contain the characteristic fur and skulls of tundra voles, but many more contain white-tipped hair, and broader skulls of some other animal. The remains of a few white feet, moreover, are unlike anything you have seen before. The claws on the third and fourth toes are very long and deep and are divided into two prongs. These are the remains of collared lemmings. The unusual claws serve as shovels for digging tunnels in the dense tundra api. Collared lemmings wear their snow shovels only in winter. The claws are replaced by more normal-looking ones when the lemmings molt from their white winter coats to their brown summer ones.

When ground squirrels awaken

Other creatures up and about on the tundra are arctic ground squirrels. Ever since your arrival, they have been squabbling noisily on a large snowbank near your cabin. Evidently a colony lives on the slope beneath the bank. Grayish-brown with several rows of rusty yellow-brown spots along their backs, they are conspicuous as they chase each other across the snow and make whistling or guttural chattering sounds.

A few weeks earlier the scene would have been silent, for the ground squirrels have just emerged from hibernation. They are, in fact, the only hibernators found in this area. There are no chipmunks, no jumping mice, nor even any deermice. These smaller hibernators probably cannot winter this far north because of the increased heat loss that would result from their greater ratios of surface area to volume.

Despite the snow, ground squirrels emerge from hibernation in early May and venture into the wintery world above the ground. Food may be scarce, but the squirrels are still fat enough to survive until the snow melts and plants begin to grow.

But how do the ground squirrels manage? You ought to be able to get some clues by digging out one of their dens. This is no easy matter. The snow in the drift is less dense than on the tundra. Step on it, and you sink in up to your thighs. As you dig down, moreover, you discover that the drift is ten feet deep. When you finally reach the entrance to the burrow in the bank, you can dig no farther. The ground is frozen solid. How can an animal keep warm in a burrow in the permafrost, where the temperature surely is below freezing?

Ecologists have studied ground squirrels in many places and have dug out many hibernating dens. They have made some interesting discoveries. For one thing, they have found that the squirrels always hibernate in areas covered by deep, fairly soft snowdrifts, which insulate the ground from the bitter cold of the arctic winter. Thus the ground never gets colder than the temperature of the deep permafrost, which here is about 10 degrees.

But even that is too cold for a hibernator. How do the squirrels keep their nest chambers warm? In the first place, even though their life processes are slowed down drastically, the hibernating squirrels still produce significant amounts of heat. Some of this heat is conserved by the insulating layers of grasses with which the squirrels line their nest chambers. Perhaps even more important is the design of the burrow itself. The entrance is always below the level of the nest chamber. Since warm air rises, heat radiated by the squirrel is trapped in the nest chamber instead of escaping through the entrance to the burrow.

Finally, the burrows are always dug into well-drained banks where the active layer is deep and the soil is relatively dry. If the soil were waterlogged, melting permafrost would flood the squirrel out as heat from the warm nest chamber escaped to the surrounding soil.

It is a good thing the burrows are warm and safe, for this is where the ground squirrels spend most of their lives. They go into hibernation in August and do not emerge again until early May. In the three or four months that they are above ground and active, they must live at a feverish pace, mating, raising litters of five to ten young, and storing up enough fat to carry them through their next nine-month-long period of hibernation.

During the brief arctic summer, ground squirrels live at a frantic pace and, by August, store up enough fat to carry them through another nine months of hibernation. When they dig burrows, the squirrels unwittingly enrich the soil by mixing and aerating it.

Jumping the season

Faced with such a brief summer, other animals also are getting a head start on the season. When they den up in the fall, grizzly bears have already mated. They do not really hibernate. Their body temperatures remain near normal, and they simply go into a deep sleep, their life processes fired by their stores of fat. Then in January or February, while still in their semidormant state, pregnant females give birth to one to four cubs.

The cubs are helpless creatures only eight and one half inches long and weighing about a pound and a half. Though covered with a coat of fine hair, they must huddle in the mother's fur to keep warm. Soon they begin nursing on her warm, nutritious milk. Feeding the young and keeping themselves warm at the same time is a drain on the mothers' energy reserves, but the big bears are equal to the task. The cubs grow rapidly and are able to keep up with their mothers when the families finally emerge from their dens to a still snowy world in April or May.

With a pair of playful cubs tagging along behind, a polar bear heads out across the jumbled surface of the polar ice cap. The cubs are born in midwinter while the semidormant mother sleeps in her den, and emerge with her when she rouses from her sleep in March or April.

Polar bears also court and mate in midsummer, long before the females head inland in the fall. In November or December, the females hole up in caves in hard-packed snowdrifts and fall off into their winter slumber. In late December or January the cubs, usually a pair, are born, and nurse on the sleeping mother's milk. Then, as early as late March, the mother rouses from her winter sleep and travels back to the coast with her playful white cubs tagging behind. They remain with her for another year or two before they are mature enough to wander off and live on their own.

While the bears are still sleeping, wolves and arctic foxes are up and about, foraging in a dark, snowy world for whatever food they can find. The foxes get the urge to mate in February, and for about a month the winter silence is interrupted by their continual barking and squalling. Then they take up residence in their dens in dry, well-drained slopes where their litters of about five pups are born anytime from late April until June. Wolf pups are born at about the same time, often in dens that the same pairs of mates have used for many years.

EARLY OWLS

The snowy owl gets a head start on the season by nesting while snow still covers the tundra. Its nest (*upper left*), built on the same site year after year, is little more than a shallow, virtually unlined depression scraped in the soil. The female alone incubates the eggs, normally six or seven to the clutch, until they begin to hatch in about thirty-two days. Meanwhile the male hunts for both, bringing lemmings and other food to his patient mate. At the lower left, the remains of a few lemmings decorate a nest where three downy owlets have just hatched.

The white down of newly hatched snowy owlets is replaced by grayish down as the birds grow older (right). Until they are old enough to fly at about sixty days, the owlets get around by jumping jerkily and flapping their wings. Below, an owl crouches protectively over her brood. Although normally silent, the snowy owl will click her bill and emit a whinnying sound if her nest is threatened.

Songs of spring

Good weather continues during your second week at Anderson River. A few snow showers one day slow the advance of spring, but south winds and constant sunlight finally raise midday temperatures above freezing. Sublimation and melting gradually thin the api and change its structure so that you now sink up to your knees with every step.

Each day brings new flights of geese and swans, and a few ducks also begin to appear from the south. The delta is no longer silent. The excited gabble of snow geese provides a backdrop for the whistling calls of swans, the cackling laughter of white-fronted geese, and the rattling cries of male willow ptarmigan. The tundra is coming to life.

As snow around the cabin melts, exposing more and more of the willows, still other songs join the chorus. The most common are the whistling notes of white-crowned sparrows. Redpolls appear out of nowhere and sing with a rippling trill. Sparrowlike Lapland longspurs fly up into the air, then descend, uttering short gushing warbles. Also flitting through the willows and calling with clear notes are a few snow buntings heading farther north to nest.

One day in late May the thinning api disappears from higher, more exposed places on the delta islands. The gab-

As nesting time nears, snow geese battle over patches of tundra where the snow has melted. The pairs of geese claim the bare areas for nesting territories and drive off all challengers or intruders.

bling of the geese now reaches a new pitch of excitement. Although the snow geese still fly in small flocks, they no longer gather in groups on the ground. Instead they are spread out over the islands in pairs that stand by patches of bare ground. Nor will they tolerate other geese that venture near their islands of bare soil. With necks outstretched and beaks open, they hiss as if to threaten the intruders. Occasionally the trespassers even provoke a furious fight, with two birds biting each other about the neck and beating with their wings.

The geese obviously are defending their bits of bare ground. Ecologists call such a defended area a *territory*. Whether the owner uses it for breeding or feeding or both, he proclaims it as his own and will defend it against all others of his own species. Fortunately for the birds' well-being, actual fights are relatively rare. Usually the owner's mere presence on his territory, advertising his ownership, or at most his threatening behavior is enough to discourage would be intruders.

Other animals have other kinds of territories and different ways of advertising or defending them. The male white-crowned sparrow flying from one high willow to another and singing at each stop is patrolling the boundaries of his territory. His song tells other male white-crowned sparrows

In late May or early June, when northern rivers begin to thaw in earnest, common eiders and many other kinds of ducks wing in from the south and settle down to breed. Although most common eiders nest on arctic coasts, a few summer in colonies as far south as Maine.

to keep out of this area where he and his mate will nest, feed, and raise their young. Male willow ptarmigans also defend territories, chasing each other back and forth across the boundaries in ridiculous-looking postures. Even the ground squirrels, with all their chasing and chattering on the snowbank, have actually been establishing territorial boundaries around their burrows.

But why is so much energy expended in establishing and maintaining territories? Although there are many kinds of territories serving different purposes, all of them accomplish two things. They spread the population out over all the available habitat, and they divide the population into "haves" and "have-nots." Where territories include feeding areas and hiding places, such as those of ptarmigan and white-crowned sparrows, the haves clearly have a much better chance of surviving than do the have-nots. But if something should happen to the haves, the have-nots are able to move in, take over their territories, and perhaps succeed in raising families.

Nesting begins

As good weather continues, the patches of bare ground enlarge each day and eventually merge along the edges of

islands and on low ridges. Pools of meltwater collect in every low spot. Long cracks appear in the deeper river channels, separating ice that floats on water from ice along the river's edge, which is still frozen to the bottom. As run-off from upstream increases, the six-foot-thick layer of floating ice is lifted up, allowing water to flow through the cracks and produce run-off channels over the ice frozen to the bottom.

When run-off channels form, ducks begin to appear in large numbers. Pintails, mallards, widgeons, and scaup fly in and land on the open water. From the northwest, over the sea ice, come flocks of eiders, oldsquaws, and black brant, small seafaring geese. Glaucous gulls, jaegers, several kinds of shore birds—the number of birds around the delta increases day by day as more and more migrants return to their nesting grounds. New flocks of swans and snow geese also appear, but, after resting a while, they rise and wing northward across the sea toward their main nesting areas on Banks Island, a couple of hundred miles farther to the north.

The birds that remain already have begun nesting. The whistling swans are adding dry grass and willow branches to mounds of dry vegetation, the remains of last year's nests. Some of them are over three feet across and more than a foot high. They probably belong to pairs that, mated for

With a series of hops across the surface, a Canada goose gradually builds up enough speed to lift its heavy body from the water. Some of these powerful fliers come to the arctic from wintering areas as far away as Florida and northern Mexico.

life, have returned to the identical nest sites for several years and continued to add new material.

The carefully woven vegetation forms a durable platform with a bowl-shaped depression in the top to contain the large creamy-white eggs. The swans lay an egg a day, covering them with down and sedges when they are off the nest, until the grassy basins contain the full complement of four or five eggs. Like other ducks, geese, and swans, however, whistling swans do not begin incubating until the full clutch is laid. As a result, all the young hatch at about the same time.

Snow geese, too, are building nests and laying eggs. Some of them look like small swans' nests, but others are little more than sedge-lined hollows scraped into frozen ground. Eggs in such meager nests may freeze, and the shells may even crack before incubation begins, but, miraculously, they hatch anyway. From the number of eggs you find, it is clear that the geese began laying within a day or two of the time when their nest sites first became free of snow.

Pairs of whistling swans, mated for life, often continue to add material to the same nest year after year. The birds' four or five dull white eggs must be incubated for thirty-five to forty days before hatching.

Good times and bad

This has been a particularly good year, with the snow melting quickly in the third week of May. By the end of the month almost all the snow geese have been able to establish territories and begin nesting. But what do you suppose happens in bad years, when the api remains as a solid blanket until mid-June? Or even a normal year, when it lasts until late May or the beginning of June?

The weather does not affect the snow geese's migration. Every year they arrive at the Anderson River delta about May fifteenth, no matter what the snow conditions may be. Apparently day length, or *photoperiod*, governs both the timing of the last stages of their migration and the maturation of eggs in their ovaries. As a result, they are ready to nest almost as soon as they arrive. They have only to wait until their nest sites are cleared of snow.

But they cannot wait indefinitely. In a bad year only birds that can establish territories on the few cleared areas are able to nest, and they lay fewer eggs than usual. The remaining eggs are broken down in the females' bodies to fill their energy needs, since their fat stores have been depleted by lack of food during the long delay. The result is a "bust" season with very few young produced.

Other animals of the far north, such as red backed voles, also are subject to bust years when spring weather prevents reproduction or kills the young. All the calves of caribou, for instance, are born at about the same time. If blizzards seriously delay the arrival of caribou on their calving grounds, or occur during calving, many of the young may die from exposure.

The impact of fluctuations in the physical environment on the breeding success of animals, in fact, is a major feature of life in the far north. Fortunately, however, northern animals have evolved various means for dealing with unpredictable conditions. The commonest solution, and the one adopted by larger animals such as swans, geese, caribou, and muskoxen, is to live a long time and breed frequently. For these animals the usual rule of nature—"Live if you can, but reproduce"—is modified in bad years to "Reproduce if you can, but live to reproduce again when conditions are more favorable."

Snow geese usually breed in loose colonies, with each pair defending the area around its down-lined nest. The most numerous of North American geese, they are eagerly sought by hunters during their fall migration.

171

MT. McKINLEY
NATIONAL PARK

The jagged crest of Mt. McKinley, perpetually capped in ice and snow, dominates the 3,030-square-mile park in central Alaska that bears its name. Reaching a height of 20,320 feet, Mt. McKinley is the highest peak in North America. This vast wilderness park offers the visitor much more than vistas of tall mountains and glaciers, however. It also encompasses huge tracts of taiga and tundra inhabited by a varied cross section of wildlife that includes grizzly bears, caribou, Dall sheep, moose, and many smaller birds and mammals. Though relatively remote, the park is accessible by automobile, train, or plane, and overnight accommodations are available at a hotel or in any of seven campgrounds.

The sleek common loon nests near lakes and ponds in Mt. McKinley National Park and across much of the North American taiga and tundra. For many people, its eerie lunatic call epitomizes the mysterious beauty of the far north.

Thick, springy carpets of moss cover much of the wet lowland tundra in the park (below left). In such places, the varied plant life includes grasses, sedges, willows, dwarf birches, blueberries, and many kinds of wildflowers. One of the most appealing is the forget-me-not, Alaska's state flower (below right).

A loud whistle piercing the silence on Mt. McKinley may mean a hoary marmot is nearby. This relative of the woodchuck lives among rock piles, often above tree line, where it raises an annual litter of three or four young.

Flocks of Dall sheep clatter across rocky slopes of Mt. McKinley National Park, the only United States national park in which they are found. In autumn the rams' dramatically curved horns often clash in battle as the males fight for possession of the ewes.

One of the earliest flowers to blossom on the slopes of arctic mountains is the perennial mountain avens, a member of the rose family. It carpets the ground with showy blooms in early summer, just after the snow melts.

Perennial plants

Much the same thing is true of tundra plants, which also must reproduce if the species is to survive. Annual plants depend entirely on seeds to carry them from one year to the next, since the parent plants die at the end of the growing season. Arctic summers are so short and the climate is often so severe, however, that there usually is not enough time to grow, flower, and set seed in a single season. Thus, if one bad season were to prevent the production of seeds, an annual species would be wiped out.

So it is not surprising that very few annuals grow in the high arctic. Instead we find species such as arctic avens in which the mature plant lives for several years. For these *perennial* plants, occasional failure to set seed in bad years is not so critical. If they fail in one year, the parent plants remain alive to produce seeds the next summer.

As further insurance against extinction, many perennials spread by other means as well. Some, for instance, produce

176

underground stems, called *rhizomes*, which radiate from the parent plant and send up new shoots some distance away. This kind of nonsexual or *vegetative reproduction* has one drawback: all the offspring are genetically identical to the parent. Thus, if conditions were to change so much that the parents could not survive, the offspring would be killed as well.

So it is better in the long run if the plant can reproduce at least some of the time by means of seeds. In this *sexual reproduction*, a sperm nucleus in a pollen grain unites with an egg nucleus in the ovule of a female plant. The resulting seed contains a combination of hereditary material from two parents and so may produce an individual better able than either of its parents to survive if conditions change.

Pollen can be transferred to the ovary in several ways. Some plants depend on the wind, but this is the riskiest means of all and is not common in the far north. Others depend on insects, which they attract with large, showy flowers and nectar-producing organs. An insect visitor in

In search of pollen and nectar, an insect pauses to alight on the clustered flower heads of a marsh fleawort. In the tundra, as in all habitats, insects play a vital role as they unwittingly transport pollen from flower to flower.

177

search of nectar inadvertently picks up some pollen, which may adhere to the sticky pollen-receiving organ of the next flower the insect visits.

Arctic flowers also attract insects in another way. They are usually shaped like concave mirrors, which reflect the light energy falling on them and focus it in the center of the bloom. As a result, the temperature at the focal point is several degrees warmer than the surrounding air. Besides helping the reproductive parts of the flower develop, this extra heat provides a warm place that is very attractive to insects.

You have probably seen the blooms of sunflowers and tulips change position during the day so that the flowers always face the sun. Arctic flowers commonly do the same thing, helping them keep their internal temperatures high but also making them continually attractive to warmth-seeking insects.

The best way of all to assure pollination, however, is to have pollen-producing and egg-producing organs on the same individual and to have eggs fertilized by pollen from the same plant. This method, called *self-pollination*, is very common in arctic plants.

Stamens heavy with pollen burst from the male catkins of an arctic pussy willow. The willows must produce a superabundance of pollen, since they depend mainly on the wind to carry the pollen from male catkins to the female catkins, which grow on separate plants.

Hatching time

Early in June the final breakup of ice on the river takes place. For some time now, the floating ice on the river has been rotting. If you kick it, it breaks up easily into long, candlelike crystals—hence the common northern name, "candled ice." Now the ice along the shore also breaks loose from the bottom. It heaves about in chunks that crush the weak floating ice, and piles up in temporary jams between islands. As water flows down from upstream, where breakup already has occurred, it eventually breaks through the jams, rushing forward in miniature tidal waves that sweep huge chunks of ice downstream or strand them on islands. Then suddenly it is over. Within eight hours the mouth of the river is transformed from solid ice to open water.

The arctic poppy's bowl-shaped blossom acts as a concave mirror, directing the sun's rays to the flower's center, where the warmth promotes development of the plant's reproductive organs and attracts pollinating insects.

179

By now the tundra is alive with birds. Phalaropes bob on the water of every pond, and sandpipers and plovers scurry along the shores. Loons and grebes are settling down to nest on low islands. Several kinds of gulls soar overhead, as do terns and jaegers.

The islands on the delta seem relatively quiet now, for the geese and swans have settled their territorial fights. The female geese incubate their eggs almost constantly, leaving their nests only for short periods to feed on nearby grasses. As a result, they lose weight rapidly, although the ganders that stand guard by the nests eat more often and remain in better condition.

Their time of hardship is not over. Sometimes late storms move in, the temperature drops, and the ground is covered by three or four inches of fresh wet snow. How well the birds fare depends on how long the bad weather lasts and what condition the geese are in. If the sun melts the snow within a day or two, most of the birds weather the storm nicely. But if bad weather lasts for several days, they may not be able to keep their nests warm and dry, especially if the beginning of the nesting season was already delayed by bad weather, and the birds were not in good condition to begin with. Then, when the nests become wet or the

Though meltwater from upstream spills over the surface, the Anderson River in northwestern Canada still is choked with ice when summer arrives. Eventually, however, the rotting ice will break free and clog the channel with grinding chunks. . . .

geese get cold, they abandon their eggs. The result is a bust year, with few young produced. The only ones that benefit are the jaegers and foxes that move in to eat the abandoned eggs.

If all goes well, however, the eggs begin to hatch in late June. The first sign of hatching is a slight crack in one of the eggs. Soon the gosling breaks a hole through the shell, and its gray beak protrudes. The imprisoned gosling peeps for a while, then pecks and struggles some more until finally it breaks out and lies there, wet, bedraggled, and exhausted.

Within a few hours the rest of the eggs in the clutch also hatch. This results in part because none of the eggs were incubated until the last one was laid, but there is more to the story. Biologists have found that the peeping of the first gosling that gets ready to hatch speeds up the development of the others, and they hatch even if they are not quite ready. When all are finally out of their shells, the female broods the young until they are dry, about an hour after the last one hatches. Then the whole family leaves the nest and walks away, the gander in the lead, the goslings following, and the goose bringing up the rear of the procession.

. . . Then floodwaters at last will surge downstream as ice jams are broken and sweep the debris away. Within hours, the scene of chaos will be over. After the breakup the Anderson River is transformed into a placid expanse of calm water.

Groups of goslings

The first few weeks are critical for goslings. As soon as they hatch, gulls arrive in search of easy meals. They hover over the geese and swoop down to try to pick off any stragglers, which they swallow whole. Arctic foxes also remain on the alert to snatch up any goslings that get separated from their parents. In some regions the birds are not even safe in the water: large pike swim up from below and gulp down the small goslings.

The geese have one defense against predators. Since all of them begin nesting at about the same time, all the young in the nesting colony hatch within a few days. When families of adults and newborn goslings leave the nesting area for better feeding grounds, they gather into loose flocks. When gulls attack, the flocks bunch up, with the ganders

Quick to dry, the snow gosling's fluffy down keeps it warm as soon as it hatches (*above*). In common with other ducks and geese, snow geese can walk and even swim almost immediately. Like the black brant family (*right*), snow geese soon abandon their nest, with the goose and gander marching watchfully at either end of the procession.

at the edges uttering threat calls to deter the marauding gulls. This in fact may be one advantage of the short hatching period: enough families are moving at the same time to form defensive flocks.

The goslings grow rapidly. At hatching they weigh about two ounces, yet within five weeks they weigh thirty times that much. If you were to dissect a dead gosling, you would find out how they manage this feat. Their gizzards and livers are so large that they alone account for half the birds' weight. With twenty-four hours of daylight for feeding, the goslings are able to eat just as fast as this impressive digestive equipment can process the food.

Because they grow so rapidly, the goslings soon become too large for most predators. Gulls, for example, rarely manage to take geese more than three weeks old. Possibly this is another advantage of the short hatching period. If

The wide-ranging horned lark breeds not only on the tundra but also on prairies, grasslands, and deserts as far south as Mexico and northern Africa. An early breeder everywhere, it frequently begins to nest while snow still covers the ground.

gulls can kill only goslings less than three weeks old and the hatch lasts one week, then gulls can prey on them for only four weeks. If the hatching period stretched out over four weeks, gulls could prey on them for a total of seven weeks, from the time the first ones hatched until the last to hatch were three weeks old. That would mean a lot more goslings killed by gulls.

An early start

Because the arctic summer is so short, it is an obvious advantage to snow geese to get a head start on the season by nesting as early as possible and to produce offspring that grow quickly. Virtually all life on the tundra, in fact, must live by this rule if the offspring are to be mature enough to survive by the time the snow returns. As we saw at Lake Hazen, redpolls had finished nesting there by late July, and both young and adults had already migrated south. Longspurs, snow buntings, horned larks, and all the shorebirds of the tundra also produce young that grow rapidly and soon are able to fly.

By midsummer the young of caribou are strong enough to keep up easily with their nomadic parents, and muskoxen calves are growing rapidly. Wolf pups are large enough to emerge from their dens by the time they are three weeks old and before long are able to accompany the adults on short excursions.

Arctic hares are perhaps the last of the larger mammals to reproduce. Their young are not born until late June or July. But unlike the young of rabbits, which are blind, almost hairless, and helpless at birth, hares are born in a much more advanced state. They are fully furred, their eyes are open, and they are soon strong enough to leave the places where they are born.

Even plants have evolved means for getting a jump on the season. As we noted in the first chapter, some of them have large food reserves in their roots, enabling them to shoot up rapidly when summer begins. One plant, arctic avens, has quite a different adaptation for ensuring an early start. It grows in cone-shaped tussocks, something like

A champion migrator, the golden plover travels to the arctic from wintering areas as far away as southern Argentina, Australia, and New Zealand. Its nest, a leaf-lined depression in tundra vegetation, usually contains four nearly undetectable mottled eggs.

Since a single collared lemming can produce several litters of as many as ten young at a time over the course of a year, it is no wonder that the little rodents periodically undergo population explosions. After building up for three or four years, however, their populations suddenly crash, with perhaps one in four hundred lemmings surviving. The causes of their regularly fluctuating population cycles remain a mystery.

squat, miniature Christmas trees. In spring, the tops of the tussocks are the first signs of plant life to be cleared of snow. As the api melts and sublimates, the thin cover of snow focuses sunlight on the tussock, which absorbs heat energy and melts the snow nearby. The result is a hollow in the snow around the tussock, which serves as a kind of miniature greenhouse in which the plant can develop very early. Soon after the snow melts, the tussocks are covered with showy, white blossoms which before long produce mature seeds.

Boom and bust

Of all tundra mammals lemmings are unique in that they alone have no single well-defined breeding season. Usually several litters are born in the course of the spring and summer months, and in some years they reproduce even in winter. Their reproductive cycles are of special interest to zoologists because lemming populations fluctuate dramatically from year to year. Sometimes the tundra seems to be alive with little rodents scurrying about on the surface of the ground, while in other years scarcely a lemming is to be seen anywhere.

Yet there is more to the story than one could observe

186

in a single summer, for their numbers increase and decrease according to a regular rhythm. A year of low population is always followed by a gradual but steady increase in lemming numbers until at the end of three or four years, the population reaches a peak. Then suddenly the lemmings are practically wiped out, only to build up in numbers again over the next few years.

In Scandinavia, these *population cycles* of lemmings sometimes result in spectacular migrations. When their numbers reach a peak, the animals stream across the land like rivers of life. Following river courses, they are gradually funneled down to the sea, where they plunge into the water and drown by the tens of thousands. Although popular fancy sometimes pictures their demise as a kind of deliberate mass suicide, the hordes of lemmings probably go to their death unintentionally. More likely they plunge into the water confident that they can swim to the far side of this obstruction in their path. But the distances prove too great, and eventually they die of exhaustion.

In North America, the natural travel routes are not as restricted, and such well-oriented migrations do not occur. Mass movements of lemmings take place about every third population peak in northern Alaska, but the lemmings move in all directions, often traveling in huge circles. The migrants are generally small, immature animals, and most of them are males, suggesting that they have been forced to wander by social pressures from the larger and better-established adults.

The mass movements begin in early spring. Even before the snow melts, many lemmings come out on the surface of the api. Their tracks always lead away from the ventilation shafts in the snow; they apparently never return to the security of the subnivean space. You may find them dead, frozen solid on the snow surface. When the snow melts, the land seems to seethe with lemmings. They run around buildings and on roads, swim across ponds and rivers, and scurry everywhere across the tundra. The remaining patches of api are riddled with their tunnels, and dead lemmings, showing no obvious signs of why they died, are scattered here and there.

At such times the tundra is alive with other species as well, for the peak lemming populations provide a heyday for predators. Snowy owls swoop down and easily snatch up more than their fill of lemmings. Pomarine jaegers, parasitic jaegers, long-tailed jaegers, glaucous gulls, and short-eared owls also gather to share in the feast. For arctic foxes, hunting is so easy that they usually succeed in raising extra-large litters of pups in peak lemming years.

Within a few weeks, the banquet comes to an end. Practically all the lemmings are gone. Perhaps one in four

hundred survives to breed again and reinitiate the cycle of increasing numbers. The lemming cycle has passed its crest; however, its effects will continue to reverberate for some time to come. The next winter, farmers in the great plains are likely to notice an unusually large number of snowy owls. The great lemming abundance of early spring means a bumper crop of owlets, but with the arrival of winter there is nothing left for them to eat. As a result, many are forced to migrate south in search of food. Possibly they never find their way back to the far north, however, for the owls too decline in numbers following the lemming "crash," as their sudden decline is usually termed. Arctic foxes also are abundant following the lemmings' peak population, but with the shortage of lemmings, they begin in desperation to come easily to baited traps. That winter, Eskimo trappers will harvest a bumper crop. Then for the next few years foxes also will be scarce.

What causes cycles?

All in all quite a few northern animals are subject to three- or four-year population cycles. In addition to lemmings, arctic foxes, snowy owls, and jaegers all illustrate this phenomenon in the tundra, while forest voles, such as red-

Like many predators, arctic foxes rely heavily on lemmings as a source of food and produce more offspring in lemming boom years. When lemming populations crash, however, the foxes follow suit by declining in numbers within a year or so.

Dwarf willows grow close to the ground, where the temperature may be as much as 15 degrees higher than it is just a few feet above the surface. Their woody stems creep across the tundra, sending up branches only a few inches high.

backed voles, act in this way in the taiga of both the Old and the New World. In contrast, a number of other species fluctuate in population cycles that last for about ten years. Snowshoe hares go through a ten-year population cycle, as do their major predators, Canada lynx. So do forest grouse, red foxes, and probably muskrats and mink throughout the North American taiga, although, curiously, these long cycles are unknown in the Old World.

In temperate climates, a few animals such as meadow mice and gray squirrels also are subject to periodic population growth, followed by a dramatic crash in numbers. In general, however, boom and bust population cycles seem to be primarily a phenomenon of the far north, where the most dramatic examples of such cycles occur.

Yet no one knows just how or why they happen. All kinds of theories have been suggested, but no one has been able to prove any of them. Destruction of their food by excessive populations of herbivores such as lemmings and snowshoe hares probably is involved to some extent, while the cycles of their predators, such as arctic foxes and Canada lynx, undoubtedly are triggered by the feast and famine situations with their prey. Cycles of animals such as ruffed grouse may be secondary ones set up when the predators run out of lemmings or snowshoe hares and turn to alternate prey. In addition, lemmings, for example, seem to breed in winter only when their numbers are on the upswing, although the cause for this is not clear. By the same token, young adults reproduce when populations are low but not when populations become dense, but not everyone agrees that social interactions have something to do with the suppression of breeding activity.

Whatever their causes, these mysterious cycles are a fascinating fact of life in the far north. But they leave many questions unanswered. Most interesting of all is the question whether cycles may be an adaptation for survival in the highly unstable environment of the far north. Possibly, for instance, they serve to release vital nutrients such as nitrogen and phosphorus, which are notably scarce in the tundra. The meager supplies are locked up in the bodies of living plants, but when the plants are eaten by lemmings, the nutrients become available again in the lemmings' droppings or in those of predators that eat the lemmings. Or perhaps the periodic expansion of lemming

190

numbers is a way of introducing the maximum amount of genetic variation in the population. In the die-off that follows, presumably only the best adapted survive. If this were to happen every three or four years, changing adaptations in lemmings would never lag far behind climatic change, which continues to go on in the far north.

In any case, the question of exactly what causes such boom and bust population cycles is an extremely interesting one to investigate. Human populations, for instance, are now increasing at a rate comparable to that which occurs during the increase phase of a lemming or hare cycle. Will we at some point fall victim to some similar natural mechanism of population control and suffer a dramatic population crash? The prospect may be frightening, but it certainly provides intriguing food for thought.

Hugging the ground

Plants, of course, are the basic source of food for lemmings, as they ultimately are for all creatures of the tundra. Like the animals, they are adapted in various ways for making the most of the short growing season of the far north. A plant like dwarf willow, which creeps almost like a vine across the ground, for instance, exposes maximum leaf surface to the sun. Conical tussocks, like those of arctic avens, also get the greatest possible benefit from the sun as it circles around the horizon during the twenty-four-hour arctic day and "night."

Even the fact that most plants are very short and live close to the ground has a definite survival value. In the high arctic, the vast majority of plants live in groups that form small cushions on the surface, with little more than bare ground between them. You can easily discover the advantage of living this way by taking the air temperature out in the open about five feet above the ground and then measuring it at the soil surface. Scientists who have done this have found that surface temperatures usually are 10 to 15 degrees warmer than they are five feet above the ground. Sometimes the difference is even as great as 30 degrees. Thus, by hugging the ground, plants live in a *microhabitat* (miniature habitat) that is much more favorable for growth than the overall environment.

Silene, or moss campion, exhibits another characteristic growth form of arctic plants. Its tight hemispherical tussocks, which capture the sun's rays from every side, trap warm air in the compact mass of leaves and stems and shield the plant's center from cold breezes.

191

The wooly insulation on the stems of lousewort, or fernweed, trap the warmth of the sun's rays and help keep the growing plant warm. The roots and flowers of some species of lousewort are edible, and Eskimo children enjoy sucking nectar from blossoms of certain varieties.

Within a cushion of plants, living conditions may be even better. Warm air trapped among leaves and stems cannot be carried away by the wind, so the temperature inside the cushion rises still higher. The centers of conical tussocks of arctic avens also stay warmer than the surrounding air. In addition, the leaves and stems of many tundra plants are covered with fine hairs. This again probably helps trap warm air and keeps it from being blown away.

Even so, a cushion of plants may live for years and never get very big, for, despite continuous or nearly continuous sunlight, the growth of arctic plants is never exactly vigorous. Since average summer temperatures are only in the forties or fifties, the warmer microhabitat at ground level still is quite cool. As a result, the life processes of plants necessarily go on at a relatively slow pace. And the growing season is short. Before long, the sun will begin to set and snow will blanket the land once more.

Basking butterflies

Insects also take advantage of the warmer temperatures at ground level. On warm sunny days, mosquitoes and black flies make life miserable by swarming constantly about your head. However, most of the insects live close to the ground where it is warmer and less windy. Even flies and mosquitoes tend to lie low on overcast days. The reason for this is that cold-blooded insects must have their muscles warmed up to certain critical temperatures before they can fly. Surprisingly, some of them manage to do this even when the general air temperature is below their critical limits for flight. The orange and black *Boloria* butterflies of the arctic, for instance, can frequently be found resting on the ground on the sunny sides of tussocks of plants. They characteristically rest with their heads pointing away from the sun and with their wings fully spread and pressed against the bare soil. This is an ideal position for getting extra heat. The insect absorbs some warmth from the soil. In addition, the incoming rays of sunlight strike its wings at an angle of nearly 90 degrees, the angle that ensures maximum absorption of incoming radiation.

When you disturb a basking butterfly, it always stays within a foot or so of the ground as it flutters off. When it lands at the base of another plant, it immediately re-

While most butterflies hold their wings upright when at rest, *Boloria* butterflies of the arctic spread their wings mothlike against the ground when they alight. By doing this, they maximize the amount of heat they can absorb from the soil and from the weak incoming rays of sunlight.

193

Despite the brevity of the arctic summer, mosquitoes and black flies flourish in astonishing numbers. Face nets, long sleeves, boots, and insect repellents all are helpful, but it is impossible to escape completely from the ever-present pests.

sumes its basking position. However, if you disturb it just as a cloud passes in front of the sun, something interesting is likely to happen. When the insect alights, it may be disoriented and settle down with its head pointing in the wrong direction. But as soon as the sun reappears, the butterfly immediately changes its position so that its head points the proper way.

Not all arctic butterflies use the same tricks as *Boloria* for getting warm. As we have already seen, the bowl-shaped blossoms of many tundra plants act as heat reflectors. Some of the smaller butterflies, as well as other kinds of insects, take advantage of this heat by basking on flowers. Some kinds of butterflies also warm themselves by tilting their bodies when they land so that their upright wings are at a right angle to the sun's rays. Finally, several kinds of arctic moths have notably heavy, dark bodies covered with fine hairs. Their dark color and large size apparently are adaptations for absorbing more heat from their environment, while the hairs serve as air traps for conserving the extra heat.

Ingenious insects

Even the life cycles of many tundra insects are especially adapted for coping with the rigors of the arctic environment. Moths and butterflies characteristically mature from egg to larva (or caterpillar) and then enter into the pupal or resting stage before transforming into full-grown adults. In temperate climates, this entire life cycle may be completed within a few weeks, and rarely requires a full year. In the arctic, however, the life cycles of many moths and butterflies are drastically prolonged. For one group of moths, the larval stage lasts as long as five or six years.

Caterpillars of these moths can suspend development at any time and remain frozen for days or even months at a time. When they thaw out again, the larvae immediately resume their normal development. No one knows exactly how they manage this feat without damage to their living tissues, but obviously it is a useful ability to possess in a climate subject to wide and unpredictable fluctuations in temperature.

Mosquitoes make life miserable for beast as well as man. Though the young caribou's thick fur provides some relief from the humming throng, its unprotected nose and eyes are at the mercy of the relentless insects.

Larvae of these species usually overwinter in the most exposed places in their habitat. Although this means that they are certain to be frozen during the winter, it also ensures that they will be warmed by the first rays of the returning summer sun and thus have the longest possible growing season.

Other insects such as mosquitoes, in contrast, must complete their life cycles in a single year. Yet they too have ways of maximizing their growing seasons. If you search carefully on the north banks of small ponds at midday, you are likely to find female mosquitoes laying eggs. Apparently they choose this time and these places because the mosquitoes are then exposed to rays of the sun coming from the south. Once again, the north banks are among the first places to be warmed when the spring sun reappears in the southern sky. Although the eggs are laid five or six inches above the summer water level, melting snow the following spring washes them into the pool, where the eggs hatch. Even though the water is cold, the larvae develop rapidly and emerge within three or four weeks as mature adults, thus completing their life cycle.

The end of summer

By late July summer on the tundra already is on the wane. At the Anderson River, the sun now begins once again to slip briefly below the horizon each night. The long summer day of continuous sunlight has drawn to a close, and in the following weeks, the nightly periods of twilight and darkness gradually grow longer.

In August, as temperatures begin to drift downward, there is a definite hint of autumn in the air. Soon frost begins to appear in the morning, and thin layers of ice cover small puddles at dawn. The leaves of cranberries, dwarf willows, and other plants gradually turn to shades of crimson, russet, and gold, briefly transforming the tundra into a mosaic of color.

As the brief arctic summer wanes, the crimson leaves of bearberry and other plants tint the tundra with a tapestry of living colors. In autumn, this creeping herb produces reddish berries that are eaten by birds and, as its name suggests, by bears.

By now, most of the small songbirds have left for warmer climates, and shorebirds are beginning to slip away on their long migration flights. Caribou herds also are drifting back toward tree line, where the animals will mate and then seek the sheltered taiga for winter. Bit by bit, the cast of characters that enlivened the tundra summer is disappearing. Only a few will remain to endure the long, cold months ahead.

Late in August or early in September snow flurries become more and more frequent. The first snowfalls disappear within hours, but eventually even the midday sun is too weak to melt the accumulating api. Now that the plants on which they feed are being buried beneath the snow, the geese are forced to desert their summer breeding grounds. Day by day, more and more flocks of snow geese pass overhead in ragged lines and disappear beyond the southern horizon. Soon only the swans and a few ducks remain, feeding along the edges of the river.

And then one day they are all gone. The tundra world is strangely silent now. For the next eight months or so, signs of life will be few and far between. Yet somehow the wolves, the muskoxen, the arctic hares, the foxes, lemmings, and weasels will manage to survive. Some of the snowy owls and ptarmigans will head south, but others will remain to eke out a living through the long bleak months of winter.

But when the sun rises again in the spring and the snow melts, the migrants will return as they have for thousands of years. Plants will grow and blossom once more, and the tundra once again will pulsate with life.

Winter comes quickly to the far north, but summer is sure to follow. Early in spring, the rough-legged hawk will return to the tundra to lay its eggs in a nest on a rocky cliff, even as ice continues to break up on the river below.

Appendix

National Parks of the Far North

It has sometimes been said that the far north needs no national parks. To a large extent it is still a trackless wilderness with little or no human population, and the climate, except during the brief summer, is inhospitable at best. The entire top of the world, some say, is already one tremendous international preserve where forests crowd darkly to the edges of the tundra, and where the animals run free and myriads of birds return each summer to nest.

In recent years, however, the arctic has been increasingly encroached upon, in countless ways and over hundreds of square miles. Military defense posts; communications stations; oil, gas, and mineral exploration sites; highways; airstrips; pipelines; hydroelectric developments; water diversion schemes—all are in the planning stages or beyond. And these are but a hint of what is to come. Already this last frontier is bracing for the onslaught.

So it is wise and proper that the nations of the far north have made a start toward setting aside tracts of land where wilderness environments can be protected, visited, studied, and enjoyed. Although national parks in the far north still are somewhat scarce, it is likely that in the future more parks, preserves, and refuges will be established, not only to accommodate conservation and scientific study but also to meet the needs of a recreation-oriented society. Canada, for instance, has recently announced but not yet officially established three new parks in the far north: Kluane National Park in southwestern Yukon Territory, Nahanni National Park on the eastern slope of the Mackenzie Mountains in the Northwest Territories, and an as yet unnamed park on Baffin Island.

On the pages that follow, attractions at some of the major national parks of the far north are summarized. For further information on United States national parks, inquiries should be addressed to the National Park Service, United States Department of the Interior, Washington, D.C. Inquiries about Canadian Parks should be addressed to the National and Historic Parks Branch, Department of Indian Affairs and Northern Development, Ottawa, Ontario. In the case of other countries, you should contact national travel information offices in their capitals or in other major cities. It should be noted that visitor access to sanctuaries in the Soviet Union is uncertain, although some Soviet preserves have been opened to visiting scientists and other tourists. Finally, if you are planning to travel in Canada,

ARCTIC LOON

you should investigate the many provincial parks scattered across the nation. Many of them are as large and as well equipped as some of the national parks.

UNITED STATES

Glacier Bay National Monument

This scenic area in southeastern Alaska is primarily a showcase for glacial activity. From the snow-capped mountains of the Fairweather Range, sixteen active glaciers creep downslope into the sea, while many smaller glaciers end inland from the coast. Wildlife is also plentiful, including black bears, brown bears, lynx, wolverines, and coyotes. Sitka deer inhabit lowland coniferous forests, while mountain goats are abundant on steep slopes, and porpoises, seals, and whales of various species live in or visit the bay waters. Loons, cormorants, waterfowl, gulls, murres, ravens, and eagles contribute to the lengthy bird list. Glacier Bay is reached by boat or plane, and there is a lodge at Bartlett Cove, the starting point for daylong boat trips in the bay.

Isle Royale National Park

This 210-square-mile park off the northern coast of Michigan's upper peninsula consists of the largest island in Lake Superior. Of all our national parks, it is the most accessible for those wishing to glimpse the life of the far north. Although areas that were burned in a forest fire in 1936 are now forested with broadleaf trees, much of this roadless wilderness is covered with true tiaga forest of white spruce and balsam fir, dotted with lakes, ponds, swamps, and bogs. The varied wildlife includes more than 200 species of birds, ranging from ospreys to warblers, but the park is best known for its resident herd of moose and the pack of wolves that prey on them.

Katmai National Monument

This remote and magnificent wilderness on the Alaska Peninsula features glacier-clad mountains, a smoking volcano, volcanic craters and lakes, and ocean bays and fiords. In the interior are forests, lakes, and the now quiescent volcanic Valley of Ten Thousand Smokes. At lower elevations, taiga woodlands clothe the slopes, while higher up are found dwarf blueberry, crowberry, and birch, and still higher is tundra. Migrating salmon and steelhead trout crowd the rivers, giant brown bears live here in numbers, and moose are common, but otherwise the wildlife is similar to that at Glacier Bay. Katmai is reached by air service to Brooks River Lodge, where there are accommodations and campgrounds. Bus trips are available to the Valley of Ten

CROWBERRY

Thousand Smokes and tours by float plane and canoe can be arranged.

Mount McKinley National Park

This spectacular park in south-central Alaska centers about the continent's highest mountain and embraces more than 3000 square miles of arctic wilderness, with elevations ranging from 400 to 20,320 feet. Aspen, birch, and cottonwood forests on the lowest slopes merge with boreal forests of white spruce at slightly higher elevations. Above tree line, at about 3000 feet, the landscape is tundra, while the mountaintops are capped with perpetual snow. Dall sheep, caribou, grizzly bears, moose, lynx, wolverines, wolves, red foxes, and many other mammals are abundant. Countless birds, including shorebirds, sparrows, buntings, and larks nest here, as do their predators, the jaegers, owls, hawks, and ravens. The park is accessible by car, train, or plane, and accommodations include McKinley Park Hotel, Camp Denali, and seven campgrounds.

CANADA

Prince Albert National Park

This 1496-square-mile park is located near the town of Prince Albert, almost exactly at the center of the province of Saskatchewan. Much of the park is boreal forest of spruce, jack pine, and tamarack, but there are scattered stands of birches, aspens, and other hardwoods, as well as many fields and meadows. The park includes hundreds of lakes and miles of trails where the visitor might encounter muskrats, beavers, moose, elk, white-tailed deer, and possibly even a wolf or a lynx. The park and nearby lodgings are accessible by road or train.

JACK PINE

Riding Mountain National Park

This 1148-square-mile park in southwestern Manitoba is on a forested plateau that rises abruptly for one thousand feet or more from the flat plain. The most common trees are black spruce, white spruce, birch, and maple. The park has the usual contingent of taiga mammals, and its seventy-five lakes attract swans, white pelicans, and cormorants. Tourist facilities include hiking and riding trails, hotels, campsites for tents and trailers, and an interpretive center.

Terra Nova National Park

Located on the north shore of the island of Newfoundland, this 152-square-mile park includes a magnificent stretch of ragged coastline with deeply indented bays and coves. Inland the

205

rocky hills are mantled with spruce forests and dotted with bogs and lakes. The forests harbor moose, black bears, lynx, beavers, woodland caribou, and birds such as bald eagles, ptarmigans, and gray jays, while gulls, murres, guillemots, and other seabirds live along the shore. There are campgrounds and cabins in the park and other accommodations in nearby villages.

Wood Buffalo National Park

This 17,300-square-mile tract of subarctic wilderness in Alberta and the Northwest Territories was set aside in 1922 as a refuge for Canada's dwindling herds of wood bison, but today the park is even more renowned as the last and only breeding ground of the whooping crane. At present about sixty of these rare birds return each summer from their winter quarters in southern Texas, and are carefully guarded by scientists and park personnel. The park is a relatively flat landscape of lake-strewn taiga, bogs, prairie, and the combined deltas of the Peace and Athabasca rivers, all of them abounding with wildlife. The park includes one campground, which is accessible by road, and a full range of services is available at Hay River and Fort Smith, which are served by modern jet aircraft. Those wishing to sample the park's wilderness by canoe or small boat can obtain supplies at Fort Chipewyan.

LONG-TAILED JAEGER

NORWAY

Börgefjell National Park

This far northern national park in Norway, about 400 square miles in area, is a remote and wild place with no roads or even footpaths, although there are two lodges for visitors. Rugged mountains, with elevations up to 5500 feet, are capped with glaciers, and the spruce and pine forests on lower slopes are dotted with lakes and streams. Mammals include moose, wolverines, lynx, hares, arctic foxes, and lemmings. Among the nesting birds are many kinds of ducks and geese, willow and rock ptarmigans, shorebirds, long-tailed jaegers, snowy owls, boreal buntings, and pipits. There are some Lapp reindeer herds in the park.

Rondane National Park

This typical Scandinavian arctic wilderness of 225 square miles is in a mountainous area in south-central Norway. Although elevations reach 7162 feet, animal life is particularly rich in the forests and meadows on lower slopes. Besides moose, reindeer, otters, foxes, and other mammals of the far north, there is a rich variety of 124 species of birds. With three tourist lodges, Rondane is better equipped for visitors than Börgefjell. Although there are no roads, some footpaths through the park are an aid to hikers and campers.

SWEDEN

Muddus National Park

Unlike the other national parks of far northern Sweden, this is essentially a lowland park rather than a mountainous one. Its 195 square miles of coniferous (mostly pine) forests include many lakes, rivers, and extensive marshes and bogs. Reindeer, moose, bears, and otters are among the larger mammals. The park is a breeding ground for the rare whooper swan, as well as for cranes, wood sandpipers, golden eagles, and ospreys. Muddus has access roads and tourist accommodations. One small area has been designated as a bird sanctuary where no visitors are permitted during the breeding season.

Padjelanta National Park

Covering an area of about 780 square miles, this is the largest and wildest of several subarctic parks in Sweden. It is a land of rugged mountain beauty, with snow-capped peaks, glaciers, rivers, valleys, and extensive coniferous forests. Animals at lower elevations include brown bears, moose, lynx, and martens, while at higher elevations there are arctic foxes, lemmings, reindeer, and one of the remaining wolverine populations in Europe. There is a single motor track into the park and a limited number of tourist lodges.

Sarek National Park

This park is adjacent to Padjelanta and the landscape is similar, with coniferous forests, woodlands of birch and dwarf willow, alpine meadows, bogs, scenic rivers and lakes, and high peaks and glaciers. Wildlife also is virtually the same, with lynx, bears, and wolverines. Although access is difficult and there are no roads or visitor accommodations, large numbers of tourists come each summer to hike and camp throughout the park.

Stora Sjöfallet National Park

This 575-square-mile park lies to the east of Sarek National Park and is perhaps the most accessible of the three mountain wilderness parks in northern Sweden. The landscape and wildlife are similar to those of the two neighboring parks although this

OTTER

207

reserve includes more rivers and lakes. There is one spectacular waterfall, but it can, unfortunately, be "turned off" at the convenience of the authorities, who use it for generation of hydroelectric power. The park can be reached by boat or road, and there are several tourist lodges. As in the neighboring parks, it is best to visit in May and June if you wish to avoid the summer crowds.

FINLAND

Lemmenjoki National Park
This second largest of Finnish national parks is located far north of the Arctic Circle near Finland's northern border. In it, the Lemmenjoki River flows through narrow gorges and plunges down waterfalls in an area of magnificent mountain scenery. Besides reindeer, wolves, bears, and other typical northern mammals, the rich birdlife includes such species as wheatears, snow buntings, horned larks, and rough-legged hawks. The park is accessible by road, and accommodations are available year-round.

ROUGH-LEGGED HAWK

Oulanka National Park
Although it is only 41 square miles in extent, this small park has notable scenic values, with deep, dark spruce and pine forests, whitewater rivers, and meadows lush with wild flowers in summer. Among the rare birds found here are whooper swans, golden eagles, and eagle owls. The park, which has facilities for visitors, can be reached by road or by air service to Rovaniemi.

Pallas-Ounastunturi National Park
With 192 square miles, this is Finland's largest national park. Located near the country's northern border, it is a scenic mountain wilderness of pine and spruce forests, lakes, streams, and extensive bogs. Mammals range in size from reindeer to lemmings, and the host of nesting birds includes rough-legged hawks, willow ptarmigans, many kinds of shorebirds, pipits, and buntings. In addition to tourist hotels the park has skiing facilities.

U.S.S.R.

Kandalaksha and Lapland Reserves
These twin reserves of 84 and 500 square miles in the northwestern corner of Russia include islands in the Barents Sea and in an arm of the White Sea and a mainland area on the Kola Peninsula. The islands in the Barents Sea are tundra, while the

other islands are forested with pines, firs, and birches. The mainland area ranges from taiga to tundra. Various parts of the reserves have a wide assortment of arctic life, including seals, reindeer, moose, bears, and many smaller mammals. The varied birdlife ranges from several species of grouse to nesting colonies of eiders and other waterbirds.

Kivach Reserve
This 40-square-mile preserve in northern Karelia near the Finnish border centers on a magnificent waterfall. The surrounding forests of firs, pines, Karelian birches, and alders are dotted with many marshes. Moose, bears, martens, foxes, ermines, mountain hares, and many other northern mammals are present, and the interesting birdlife includes blackcocks and hazel grouse. The waterfall is a tourist attraction, but, like the Kandalaksha and Lapland Reserves, this one is also a site for scientific studies of the taiga.

Pechora-Ilych Reserve
This vast park of 2774 square miles in the northern Ural Mountains includes the upper reaches of the Pechora River and its tributary, the Ilych, which are important spawning areas for salmon. Most of the reserve is on lowlands, clothed either in open pine forest or in dense, dark spruce taiga. On the western slopes of the Urals the forests become stunted and then give way to mountain tundra. Beavers, moose, brown bears, reindeer, and even Siberian sable find sanctuary in the park, which also includes an experimental station for the domestication and breeding of moose. Much of the park will be submerged if the Soviet government goes ahead with a gigantic plan to reverse the flow of the Pechora and other northern rivers to supply water to southern areas of the country.

SIBERIAN SABLE

Northern Life in Mountainous National Parks

MOUNTAIN GOAT

Although most national parks of the far north are too far away to be easily visited, it is possible to experience similar habitats closer to home. This is so because climate grows colder at higher altitudes as well as with increasing distance from the equator. As a result, in many places in North America you can take a sort of compressed journey to the "far north" by simply climbing a mountain. In such places you can travel in a single day from broadleaf forests, through coniferous forests, and beyond tree line to mountaintop tundra. Because this tundra results from altitude rather than latitude, it is called *alpine tundra*, but it is genuine tundra. In the United States, mountaintop "islands of the far north" are especially common in the Cascades, Sierras, and Rocky Mountains, but even the relatively low Adirondacks in New York and White Mountains in New Hampshire have isolated summits that extend beyond tree line.

Although the coniferous trees in western mountain forests are different species from those found farther north, the forests are similar to the taiga in many respects. Beyond tree line, however, many of the plants and some of the animals are very closely related to those found in boreal regions. They got to the mountaintops as a result of the ice age. When glaciers covered much of North America, boreal plants and animals lived far south of their present ranges. As the glaciers retreated and the climate grew warmer, the boreal plants and animals gradually died out in southern areas as their ranges receded northward with the ebbing glaciers. However, some of the boreal plants and animals found cold climates and suitable living conditions on mountaintops, and there they have remained to the present, stranded on boreal "islands" in temperate regions.

Described below are notable features of some of the mountainous national parks in the United States and Canada where it is possible to observe far northern wildlife without traveling all the way to the far north.

210

UNITED STATES

Glacier National Park *(Montana)*
This magnificent park on the Canadian border adjoins Canada's Waterton Lakes National Park and the two together form the Waterton-Glacier International Peace Park. High mountains, glaciers, waterfalls, lakes, extensive coniferous forests, and alpine tundra are all found here, with well-maintained trails, roads, camps, and other facilities. Among the fifty-seven species of mammals in the park are mountain goats, bighorn sheep, moose, elk, mountain lions, and northern bog lemmings. Birds are also plentiful, including such mountaintop tundra dwellers as rosy finches and water pipits.

Grand Teton National Park *(Wyoming)*
The Teton Range, with elevations reaching almost 14,000 feet, is truly a scenic treasure. The jagged peaks have permanent snowfields, glaciers, tundra, and, at lower elevations, forests of lodgepole pine, Engelmann spruce, alpine fir, and aspen. Among the mammals are bighorn sheep, elk, grizzly and black bears, marmots, and pikas. On alpine meadows you may see mountain bluebirds, and on Jackson Lake you are likely to find trumpeter swans and many ducks. The park is easily accessible and has excellent facilities for visitors.

BIGHORN SHEEP

Mount Rainier National Park *(Washington)*
Mount Rainier, a 14,410-foot-high extinct volcano, is truly an arctic island in a temperate zone. The lower slopes are cloaked with mixed coniferous forests, while beyond tree line (at about 5500 feet) there are expanses of tundra and alpine meadows that are carpeted with wild flowers in summer. But most of the mountain is capped with permanent snowfields that feed a host of glaciers. Wildlife is similar to that of other western mountain parks, with bears and mountain goats particularly evident. The park is splendidly equipped for visitors, with roads, trails, information centers, campgrounds, and inns.

North Cascades National Park *(Washington)*
This pristine park just south of the Canadian border was established in 1968 to preserve a scenic wilderness of rugged canyons and jagged peaks with more than 150 glaciers. Mountain goats, mountain lions, grizzly bears, and black bears are among the varied wildlife to be seen here. Although the park has only one access road, there are extensive hiking trails and camping facilities.

Olympic National Park (Washington)

This giant park on Washington's Olympic Peninsula is a varied wilderness that ascends from the rocky Pacific coastline through temperate rain forests to alpine meadows, glaciers, and jagged mountain peaks. The fifty-six species of mammals in the park include mountain goats, snowshoe hares, elk, and Olympic marmots, and birdlife is even more varied. As in most national parks, there are campgrounds, cabins, and lodges, while private hotel accommodations are available outside the park boundaries. A good road enables visitors to drive to alpine surroundings on Hurricane Ridge and at Deer Park.

Rocky Mountain National Park (Colorado)

This scenic area of the Rocky Mountains includes peaks as high as 14,000 feet. Traveling up the mountains, you pass through forests of ponderosa pines and then through stands of spruce and fir. Beyond tree line are snowfields and expanses of alpine tundra which are easily accessible by car on Trail Ridge Road. Here you may see rosy finches, juncos, white-tailed ptarmigans, and ravens. In roadside rock piles you will find yellow-bellied marmots and pikas, while in the distance you may spot bighorn sheep and other large mammals. Look also for the blue columbine, Colorado's state flower.

Sequoia and Kings Canyon National Parks (California)

These adjoining parks in the Sierra Nevada are most famous for their groves of giant sequoias and the vistas of Mount Whitney, whose 14,495-foot summit is the highest point in the United States south of Alaska. Glacier-carved scenery, dense coniferous forests, and, beyond tree line, extensive alpine meadows beneath snow-capped peaks are other attractions. Hundreds of miles of trails crisscross the park, enabling the ambitious backpacker to see authentic mountain wilderness within easy range of the heavily populated West Coast.

Yellowstone National Park (Wyoming, Idaho, Montana)

America's oldest national park is most famous for its geysers and hot springs but also includes large areas of mountain wilderness, high-altitude meadows, and alpine tundra. At high elevations visitors may see bighorn sheep, mountain goats, elk, and grizzly bears, and at lower elevations there are bison, moose, pronghorn antelope, and mule deer. Among the breeding birds are the rare trumpeter swan and the white pelican. The park is well equipped with comfortable accommodations, but it is also possible to travel on foot or on horseback to remote areas where you can see one of the nation's great unspoiled wildernesses.

BLUE COLUMBINE

212

Yosemite National Park (California)

This scenic park in the Sierra Nevada centers on the beautiful glacier-carved Yosemite Valley, with the plumelike Yosemite Falls spilling down one of the rocky cliffs that wall the valley. With altitudes up to 13,000 feet, however, the park offers ample opportunity to sample alpine habitats. The rugged mountain country with broad, flower-decked alpine meadows and scores of lakes and streams is a paradise for the hiker who wants to escape the pressures of populated areas.

CANADA

Banff National Park (Alberta)

Canada's oldest national park, located on the eastern slope of the Canadian Rockies, includes 2564 square miles of scenic mountain landscapes where the visitor can observe life zones ranging from perpetual snow to alpine tundra, coniferous forests, and wooded valleys. Spectacular glacial lakes are scattered throughout the park. The higher elevations have mountain goats, bighorn sheep, pikas, ravens, golden eagles, pipits, and rosy finches, while in the forests there are moose, elk, grizzly and black bears, mule deer, martens, lynx, and many smaller animals. Especially well equipped with tourist facilities, including elaborate resorts at Banff and Lake Louise, this is one of Canada's most popular parks and is often crowded in summer.

Glacier National Park (British Columbia)

This 521-square-mile park high in the Selkirk Range has peaks towering to more than 11,000 feet, rushing streams, alpine tundra, forested valleys, canyons, caves, and more than 100 glaciers. As in other mountain parks, the tundra is carpeted with mosses, flowering plants, and dwarf shrubs. Because of the depth of winter snows, hoofed animals are rare in the park, but grizzly bears are often seen on the numerous avalanche slopes searching for edible plants or digging out ground squirrels.

Jasper National Park (Alberta)

With 4200 square miles, this is the largest of Canada's mountain parks. It is especially noted for dramatic mountain scenery, with snow-capped peaks and extensive icefields and glaciers. In summer the alpine tundra blooms with phlox, moss campion, heaths, and arctic willow, while lower meadows blaze with lupines, arnica, ragwort, mountain daisy, glacier lily, and Indian paintbrush. Wildlife here and in the evergreen forests below tree line

PIKA

213

is similar to that at Banff, with the addition of a resident wolf population. The park has a splendid variety of tourist facilities.

Kootenay National Park *(British Columbia)*
Located just south of Yoho National Park, and adjacent to Banff, Kootenay features high peaks, glaciers, and deep valleys, but is also known for its hot springs, spectacular canyons, and other unusual geological formations. The park's 543 square miles are laced with trails for riding and hiking, and there are lodges, cabins, and campgrounds as well as accommodations in nearby communities.

Yoho National Park *(British Columbia)*
This 507-square-mile park on the western slope of the Canadian Rockies is one of the most beautiful of Canada's mountain parks, with icefields, glaciers, forested slopes and valleys, and serene mountain lakes. The most spectacular of several waterfalls in the park is Takakaw, which plummets down for 1800 feet. Wild-life and vegetation are similar to that in nearby Banff and Kootenay National Parks, and the park has campsites, trails, and picnic areas.

GLACIER LILY

Man and Permafrost

One of the inescapable facts of life in the far north is ground that remains permanently frozen, except for a thin surface layer that thaws in summer. This is the famous—and notorious—permafrost, the perennially frozen ground that underlies most of the arctic lands of the world. During the colder months, it provides a firm and solid footing for man's highways, making winter travel feasible in areas of level landscape and poor drainage, where the terrain would be an impassable quagmire or bog in summer.

But permafrost may exact penalties for its use. Unless it is understood and reckoned with, it can lure men into costly failures that result in environmental havoc. Ignorance, carelessness, greed, or haste can easily destroy the delicately balanced arctic environment, and leave scars that take decades to heal—if they ever heal at all.

The price we pay

The most important feature of permafrost is its water content. If permafrost contains much water in the form of ice, it remains solid only as long as its temperature remains below freezing. Exposing it to solar radiation, or to the warming effect of heated buildings, or to summer air, or to running water, or to heated crude oil, or to ponded water causes it to thaw from the surface down.

When this happens, the result is predictable. As the ice in the soil melts, soil particles settle out and the water runs off, evaporates, or collects in pools at the surface. Worse still, the spaces in the soil that were occupied by ice crystals are now empty. As a result, the soil settles, compresses, shifts, or collapses in a process called thermokarsting, and whatever has been built upon it tilts, sinks, or cracks. This has been a crucial problem for construction in the arctic, and engineers now realize that if they are to build upon permafrost, it must be shielded from thawing.

The price we must pay for disregarding this simple fact of life in the arctic is obvious to anyone who flies a plane over the tundra. Around the sites of abandoned camps and even in seemingly unvisited wilderness, the landscape is defiled by the tracks of long-departed vehicles. By tearing up the fragile insulating layer of tundra vegetation, passing vehicles lay bare the soil

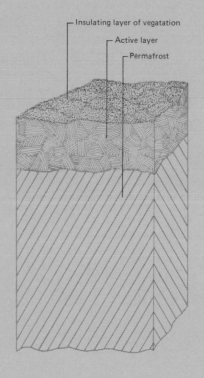

Insulating layer of vegatation

Active layer

Permafrost

and permit it to thaw into watery scars that, like footprints on the moon, may remain for tens or even hundreds of years. Merely crushing the vegetation may increase the penetration of solar radiation and result in slumping of the surface.

Today it is difficult to find sizable tracts, even in the remotest areas, where man's ugly wounds do not show. The land along seismic transects and around electronic warning systems, weather stations, trading posts, settlements, and military outposts everywhere is scarred by vehicle tracks, the indelible human graffiti on the pristine landscape of the arctic. And around most human settlements, the land is littered with abandoned buildings and discarded materials that were too difficult to bury in the frozen soil and not worth the cost of transporting to the "outside."

Building roads and highways

Unless they are correctly built, highways and airstrips create the same sorts of scars, but on a much larger scale. In the first excitement over the discovery of a large oil field at Prudhoe Bay on Alaska's North Slope, a highway was bulldozed straight across Alaska for 450 miles without regard for soil or vegetation. During the first winter, the road remained frozen and usable. But when summer came, the stripped earth was warmed by the sun and parts of the highway soon came to resemble a canal that could be traversed almost as easily by canoe as by truck.

Yet as the Canadians, Alaskans, and Soviets have proved, it is possible to build highways across permafrost. A prime consideration is careful site selection. Once again, ice content is the key. It is best to build on bedrock or coarse, well-drained soil with low ice content. By sticking to these kinds of terrain, it is possible to avoid settling. Fine-grained or peaty soils with high ice content, on the other hand, are certain to settle and shift when thawed. The most hazardous of all are soils with ice wedges or ice lenses that leave large cavities when they thaw.

Thus, highways can be safely built if they are routed over well-drained soils and bedrock and if supersaturated fine-grained and peat permafrost are avoided. In addition, it is necessary to avoid cutting into or through hillsides, since this causes the exposed permafrost to thaw and gives rise to slumping and mud slides. Where high-ice-content soil cannot be avoided, the permafrost must be protected by covering it with thick, well-maintained insulating layers of gravel.

By observing all these precautions, highways and airstrips can be and have been built with minimal damage. Yet the massive use of gravel insulation for road building and other types of construction poses the possibility of another environmental prob-

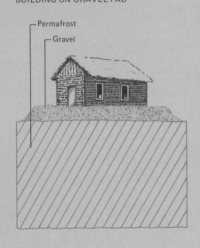

BUILDING ON GRAVEL PAD

Permafrost

Gravel

lem. In the arctic, gravel is found mostly in broad shallow riverbeds and along beaches. Many people are concerned at the prospect that removal of the millions of cubic yards of gravel that would be needed for major exploitation of the arctic would destroy the rivers as spawning grounds for fish and habitat for other animals.

Housing hazards

In building homes and other structures, the key to maintaining stable soil again is some form of insulation, for if the underlying permafrost thaws, the building may settle and tilt. For small lightweight structures, an airspace between the floor and the soil surface is often enough to protect the permafrost. This can be accomplished by placing the building on cribbing or stilts, or it can be built on a gravel pad four or five feet thick. Larger structures can be raised on pilings driven deep into the permafrost, with an airspace or a gravel pad beneath the building or an arrangement for circulating refrigerated air between the building and the ground. Using these techniques, the Russians have built several entire cities on permafrost in Siberia. Many of the larger structures rest on concrete pilings that penetrate as much as sixty feet into the permafrost, with airspaces between the buildings and the ground.

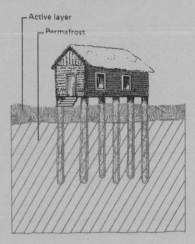

BUILDING ON PILINGS

Active layer

Permafrost

Establishing a town in the arctic involves more than simply erecting buildings, however. Even providing adequate water supplies can be a problem. Where permafrost is hundreds of feet thick, it is impossible to sink wells through it to tap any water that may be underneath. Using surface water can be equally difficult, since the ponds or reservoirs may freeze to the bottom in winter. As a result arctic communities must rely on melted blocks of ice for water supplies.

Sewage disposal is an even greater problem since the warm waste waters magnify the thawing problems when they are piped away. And because the cold climate impedes bacterial action, the wastes decompose so slowly that odors persist. At present, the favored technique is to build sewage lagoons or diked ponds where the sewage can decompose. Like reservoirs, these impoundments require constant maintenance in summer because their warmth tends to thaw the surrounding permafrost and cause slumping of the banks and dikes.

Even the transportation of water and sewage poses problems. If pipelines are placed underground, the warmth of the fluids they contain will thaw the permafrost, causing the pipeline to float in some sections and settle in others. The resulting strains could cause breaks and leaks. If the pipelines are built on the surface, however, the contents will freeze and break the pipes. Some arctic communities, such as Inuvik in the Northwest Territories, have solved the problem in a unique way. All public utility lines are enclosed in insulated, aboveground conduits called utilidors. The water in both hot and cold lines circulates constantly; heat escaping from the hot water lines keeps the temperature in the utilidors above freezing. Although utilidors are expensive to build and maintain, they are the best solution yet devised.

The pipeline problem

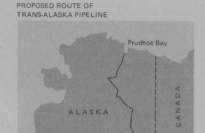

PROPOSED ROUTE OF
TRANS-ALASKA PIPELINE

Similar but even greater hazards are involved in the construction of oil pipelines such as the proposed Trans-Alaska pipeline which would carry oil from the wells at Prudhoe Bay on the Arctic Ocean to waiting tankers at Valdez on Alaska's south coast. The problems result because, since the interior of the earth is hot, crude oil is hot when pumped to the surface. The temperature of the oil at Prudhoe Bay is about 160 to 180 degrees Fahrenheit. As this hot oil comes to the surface, it loses some of its heat to its surroundings. To prevent this heat from thawing the permafrost around the well casing and causing a blowout, engineers have devised a system of refrigerating the upper part of the casing. It is also technically feasible to erect large storage tanks for hot oil on permafrost. But the real problem remains unsolved: how do you transport hot fluid through a pipeline laid in or on soil with a high ice content?

Suppose, for example, that the pipeline were laid directly above a long ice wedge. After the wedge melted, a long span of the pipe would be left unsupported and subjected to severe strain. Several ways to overcome these problems have been considered. Cooling the oil, although it seems obvious, is not a practical solution, since cold oil does not flow as easily as warm oil. In any case, the oil would soon be heated by the energy imparted to it by the gigantic pumps needed to keep the oil moving. Raising the pipe off the ground on trestles, on the other hand, would create a real barrier to migrating caribou and possibly other large animals. Placing the pipeline on a thick gravel pad and covering it with more gravel, producing a man-made "esker," would cause less of a barrier to migrating animals but would require many millions of cubic yards of gravel.

This pipeline in particular involves one further very special danger. It must cross three active earthquake faults including the massive Denali Fault. The spector of earthquake damage to a pipeline containing almost half a million gallons of oil per mile has been an important deterrent to final approval of this controversial project. If the pipeline is finally built, it is obvious that many precautions will be necessary if the surrounding landscape is not to be destroyed.

Farming in the far north

Attempts to cultivate arctic land are also affected by permafrost. Crops can be planted only where the active layer is deep enough to provide nourishment for crops and the soil becomes warm enough in summer for seeds to germinate and plants to grow. The soil generally is poor in nutrients and requires heavy doses of imported fertilizers. Because of these factors and the short growing season, relatively few crops can be grown successfully. The best results have been with vegetables such as potatoes, cabbage, radishes, and turnips, but the record so far has been discouraging. Despite hard work and constant care, harvests are usually meager. However, little effort has been directed to cultivating plants already adapted to northern conditions such as the several kinds of berries that grow in profusion on the tundra.

As a result, agriculture in the North American far north has so far been confined mainly to small dooryard vegetable plots, and has been restricted almost entirely to the taiga. Even here, stripping away the forest may result in thawing of the permafrost, settling of the land, and creation of serious drainage problems. If the permafrost hides any sizable blocks or wedges of soil with a high ice content, the soil collapses into deep water-filled pits or bogs when the ice melts. Because of all these problems, most attempts at farming on permafrost produce only marginal success, and after a few years, the scarred and damaged land is abandoned.

All in all, it is clear that man can live with permafrost only if he learns to respect—and protect—it. To ignore it or to try to bulldoze the problems aside is to court financial ruin and environmental disaster.

THERMOKARST CAUSED BY
MELTING OF ICE LENS

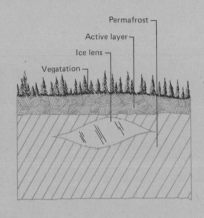

Permafrost
Active layer
Ice lens
Vegatation

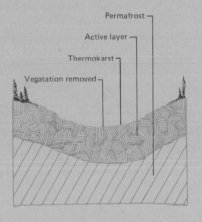

Permafrost
Active layer
Thermokarst
Vegatation removed

Vanishing Wildlife of the Far North

For all creatures that live there, the far north is a perilous habitat; yet, down through the centuries, most arctic animals have managed to survive and even flourish in their harsh environment. In historic times, however, human interference has become a real threat to the survival of many species. As early as 1768, explorers already had killed off the last of the world's Steller's sea cows. By the turn of this century, Alaska fur seals and northern sea otters also had been practically exterminated for the sake of their pelts, while whalers were all but eliminating several species of whales from arctic waters. Even sea birds fell victim to human greed: the world's last great auk was killed in 1844 and the last Labrador duck seen alive was captured in 1875.

The slaughter still goes on. Eskimos who once used primitive weapons are now equipped with guns and snowmobiles. Wealthy "sportsmen" hunt from helicopters for polar bears and other trophy animals. Oil-rich sheiks and rajahs pay as much as

HUDSONIAN GODWIT

This large and handsome shorebird is also one of our rarest and most elusive. During the nineteenth century it was so heavily hunted for meat as it migrated between northern Canada and southernmost South America that, by 1920, it seemed doomed to extinction. Since then, however, the bird has gradually increased in numbers as a result of the discontinuation of market hunting and legal protection in most countries along its migration route.

POLAR BEAR

Only fifty years ago, the polar bear was fairly numerous along all arctic coastlines, but in the last half century its population has declined drastically. No one knows exactly how many bears are still alive, but most scientists believe that fewer than 10,000 remain. The major cause has been excessive hunting, both by Eskimos who supply the fur trade and by trophy hunters who track bears by plane. Unless more stringent protective measures are enacted, the polar bear's future appears bleak.

$25,000 each for white gyrfalcons, the royal birds of falconry. And trappers continue to take their toll of martens, sables, and other fur-bearers. In the long run, however, increasing habitat destruction by humans is likely to prove an even more potent threat to the existence of far northern wildlife.

Fortunately the picture is not completely gloomy. As the public's ecological awareness increases, steps are being taken to protect arctic wildlife, and more and more research is being undertaken. Thanks to stringent international controls, for example, Alaska fur seals no longer are threatened with extinction, and strict protection of a small population of sea otters in California is resulting in an increase in their numbers. More and more parks and refuges also are being established, to the great benefit of species such as wood bison, whooping cranes, and muskoxen. Ontario has even established Polar Bear Provincial Park on the shores of Hudson Bay as a refuge for those beleaguered beasts.

Although these measures bode well for the future prospects of some species, the outlook for others is not at all hopeful.

RIBBON SEAL
This handsomely marked seal of north Pacific and arctic seas never was abundant anywhere, and its declining population now is estimated by some to be as low as 5000 individuals. Despite the low commercial value of its fur, however, limited numbers still are taken each year by Japanese sealers and Alaskan Eskimos. And in Alaska it is a victim of a government bounty paid on the pelts of all seals in an attempt to protect the salmon fishing industry from predation by seals.

ATLANTIC WALRUS
This rare subspecies of the walrus continues to decline throughout its range in the North Atlantic from North America to northern Europe and in some places is all but extinct. Excessive hunting, primarily by Eskimos, is the main cause of the decline. With an annual kill of over 10 percent of a population estimated in 1966 to number 25,000, continued depletion seems likely. Several countries now limit or forbid its hunting, but the effectiveness of these regulations remains doubtful.

Glossary

Active layer: The surface layer of soil above *permafrost* that thaws each summer and freezes each winter.

Adaptation: An inherited structural, functional, or behavioral characteristic that improves an organism's chances for survival in a particular *habitat*.

Alpine tundra: *Tundra* found beyond *tree line* on high mountains.

Annual: A plant that completes its life cycle from seedling to mature seed-bearing plant during a single growing season and then dies. *See also* Perennial.

Api: Alaskan Eskimo word, now widely used by ecologists, for snow lying on the ground. *See also* Qali; Qamaniq.

Arctic: Pertaining to the region of the earth north of the *Arctic Circle*.

Arctic Circle: An imaginary line parallel to the *equator*, circling the earth at 66 degrees 30 minutes north *latitude*; the southern boundary of the arctic region.

Aurora borealis: Display of lights of varying colors, intensities, and patterns in northern skies, caused by collisions of charged particles radiating from the sun with rarefied gases high in the earth's atmosphere.

Boreal: Northern; of a northern character.

Boreal forest: Northern coniferous forest; *taiga*.

Boreal woodland: Open, parklike subzone of the *taiga* just south of *tree line*, where the widely spaced trees do not form a continuous canopy.

Carnivore: An animal such as a wolf or lynx that lives by eating the flesh of other animals. *See also* Herbivore.

Catkin: A scaly spike of small clustered flowers that grows on willows, birches, and certain other trees.

Circumpolar: Surrounding one of the poles. A plant or animal species that lives in high-latitude areas throughout the Northern or Southern Hemisphere is said to have a circumpolar distribution.

Climate: The average long-term weather conditions of an area, based on records kept for many years and including temperature, rainfall, humidity, windiness, and hours of sunlight.

Cold-blooded: Lacking the ability to regulate body temperature, with the result that body temperature fluctuates substantially with variations in external temperature. *See also* Warm-blooded.

Competition: The struggle between individuals or groups of living things for such common necessities as food or living space.

Conduction: The transfer of heat energy from one molecule to another with which it is in contact.

Coniferous: Cone-bearing; referring to a plant that bears its seeds in cones. The term usually refers to needle-leaf trees such as pine, spruce, and fir, although some coniferous species, such as yew and juniper, bear fruits that look like berries rather than cones.

Conservation: The use of the earth's natural resources in a way that ensures their continuing availability to future generations; the wise use of natural resources.

Crustose: Crustlike; used with reference to those types of *lichens* that grow in a thin crust adhering firmly to rocks or soil.

222

Deciduous: Describing a plant that periodically loses all its leaves, usually in autumn. Most North American broadleaf trees are deciduous, as are a few *coniferous* trees such as tamarack and cypress. *See also* Evergreen.

Ecologist: A person who studies ecology, the science which analyzes the relationships of living things to each other and to their nonliving environment.

Environment: All the external conditions, such as soil, water, air, and organisms, surrounding a living thing.

Equator: An imaginary circle around the middle of the earth, equally distant at all points from both the North and South Poles.

Esker: A serpentine ridge of gravelly or sandy soil deposited where a stream flowed beneath or through a *glacier*.

Evergreen: A plant that does not lose all its leaves at one time. Most North American *coniferous* trees are evergreen. *See also* Deciduous.

Foliose: Leaflike; when used to describe *lichens,* refers to those types that grow in loose leaflike forms attached to rocks or other objects.

Fruticose: Shrublike; when used to describe *lichens,* refers to upright branching forms such as caribou lichen or reindeer moss.

Glacier: A large mass of ice that forms on high ground wherever winter snowfall exceeds summer melting. As snow and ice continue to accumulate at its center, the mass moves slowly downslope until it melts or breaks up.

Habitat: The immediate surroundings (living place) of a plant or animal; everything necessary to life in a particular location except the life itself.

Herbivore: An animal that eats plants. *See also* Carnivore.

Hibernation: A prolonged dormant or sleeplike state that enables an animal to escape the difficulties of survival during winter months in a cold *climate*. Nearly all *cold-blooded* animals and a few *warm-blooded* animals hibernate during the winter in cold climates.

Ice floe: A floating sheet of ice.

Ice-wedge polygons: A type of patterned tundra soil in which the surface is broken up into many-sided islands separated by shallow troughs; results from the contraction and cracking of surface soil when it freezes, permitting water to enter the cracks and freeze into wedge-shaped veins of ice.

Insulation: A protective covering of a substance that is a poor conductor of heat.

Latitude: Distance north or south of the *equator,* measured in degrees.

Lichen: Any of a large group of plants consisting of an alga and a fungus living in such close association that they appear to be a single plant.

Microhabitat: A miniature *habitat* within a larger one; a restricted area, such as in the air space under the snow, where environmental conditions differ from those in the surrounding area.

Midden: A refuse heap, such as the piles of pine or spruce cones deposited by red squirrels at their feeding stations.

Migration: A periodic, especially seasonal or annual, movement from one place to another of large numbers of a species of animal.

Mutualism: The form of *symbiosis* in which both partners benefit from the relationship.

Northern lights: The *aurora borealis.*

Perennial: A plant that lives for several years and usually produces seeds each year. *See also* Annual.

Permafrost: A layer of permanently frozen soil and other deposits, sometimes as much as two thousand feet thick, found in regions where the average temperature for the whole year is below freezing.

Permafrost table: The boundary between *permafrost* and the *active layer* above it; the maximum depth to which the active layer thaws in summer.

Photoperiod: The length of time during which a plant or animal is exposed to light each day, considered especially in terms of the effect of light on growth and development.

Pollination: The transfer of pollen from the male to the female organs of flowers, resulting in the formation of seeds.

Population cycles: Rhythmic fluctuations in the abundance of certain animals, in which periodic increases in their numbers are followed by sharp and sudden declines. Although many animal species undergo regular population cycles, the cycles usually are most pronounced in tundra and desert areas.

Predator: An animal such as a wolf or fox that lives by capturing other animals for food. *See also* Prey.

Prey: An animal that is captured for food by another animal. *See also* Predator.

Qali: Alaskan Eskimo word, now widely used by ecologists, for snow caught on the leaves and branches of trees. *See also* Api; Qamaniq.

Qamaniq: Alaskan Eskimo word, now widely used by ecologists, for the bare or relatively bare space on the ground beneath snow-covered trees. *See also* Api; Qali.

Radiation: The transfer through space of energy, such as heat or light, in the form of rays or waves.

Radiation shield: Anything that blocks the passage of radiating energy waves.

Rhizome: A horizontal rootlike stem that grows in the soil.

Self-pollination: Fertilization of a plant's egg cells by pollen produced on the same plant. *See also* Pollination.

Sexual reproduction: Formation of a new generation of a plant or animal through the union of male and female germ cells.

Solifluction: "Soil flow"; the downhill movement of soil and rock as a result of weather conditions, such as the downhill flow of muddy soil of the *active layer* above *permafrost* when it thaws in summer.

Species (singular or plural): A group of plants or animals with many characteristics in common. Individuals belonging to the same species resemble each other more closely than they resemble individuals of any other species and usually interbreed only with each other.

Sublimation: The conversion of solid matter, such as ice, directly into the gaseous state, such as water vapor, without passing through the intermediate liquid state.

Subnivean: Located or occurring under the snow.

Subspecies (singular or plural): A group of individuals of the same *species* living within a more or less well-defined geographical area and differing slightly but consistently from individuals of the same species living elsewhere. A single species of plant or animal may include many subspecies.

Succession: The process of continuous, gradual replacement of one community of plants and animals by another over a period of time, such as the change from bare field to mature forest; leads eventually to a more or less stable association of the living things best suited for survival under existing conditions of soil, climate, and other environmental factors.

Symbiosis: An association of two dissimilar organisms in a relationship that may benefit one or both partners. In the case of the symbiotic relationship called parasitism, however, one partner (the parasite) benefits, but its counterpart (the host) is actually harmed by the association.

Taiga: *Boreal forest*; the circumpolar forest of high latitudes composed primarily of *coniferous* trees such as spruces but also including large tracts of *deciduous* trees such as aspens, birches, and willows.

Territory: An area defended by an animal against others of the same species. A territory may be used for breeding, feeding, or both.

Till: Mixed, unsorted soil, gravel, and rocks deposited by a *glacier*.

Tree line: The farthest limit of tree growth in northern regions and on mountains; the line beyond which living conditions are too severe to permit the growth of trees. It is also called timber line.

Tundra: A *habitat* characterized by short annual growing seasons, severe winters, low annual precipitation, and an absence of trees. Tundra is found primarily beyond *tree line* in northern regions, where it is called *arctic* tundra; it also occurs above tree line on high mountains, where it is called *alpine tundra*.

Vegetative reproduction: Formation of a new generation of plant or animal by nonsexual means, such as the production of runners on a strawberry plant. *See also* Sexual reproduction.

Warm-blooded: Able to maintain a fairly constant body temperature in spite of fluctuations in environmental temperature. Of all animals, only birds and mammals are warm-blooded. *See also* Cold-blooded.

Wilderness: A tract of land, whether tundra, forest, seashore, desert, or any other, where man is only a visitor; an area where the original natural community of plants and animals survives in balance and intact, unaltered by mechanized civilization.

Yard: A sheltered area where large numbers of deer, moose, or other antlered animals gather in winter.

Bibliography

THE NORTHERN WORLD
BAIRD, PATRICK D. *The Polar World.* Wiley, 1964.
CARRIGHAR, SALLY. *Icebound Summer.* Knopf, 1953.
CRISLER, LOIS. *Arctic Wild.* Harper & Row, 1958.
DYSON, JAMES L. *The World of Ice.* Knopf, 1962.
FREUCHEN, PETER, and FINN SALOMONSEN. *The Arctic Year.* Putnam, 1958.
HANSEN, HENRY P. (Editor). *Arctic Biology.* Oregon State University Press, 1967.
LEY, WILLY, and THE EDITORS OF LIFE. *The Poles.* Time, Inc., 1962.
SETON, ERNEST THOMPSON. *The Arctic Prairies.* Scribner, 1911.
STANWELL-FLETCHER, THEODORA M. C. *The Tundra World.* Little, Brown, 1952.
SUTTON, GEORGE MIKSCH. *High Arctic.* Erikson, 1971.

ARCTIC WILDLIFE
LEOPOLD, ALDO STARKER, and F. FRASER DARLING. *Wildlife in Alaska.* Ronald Press, 1953.
MURIE, ADOLPH. *A Naturalist in Alaska.* Doubleday, 1963.
PRUITT, WILLIAM O., JR. *Animals of the North.* Harper & Row, 1967.
STONEHOUSE, BERNARD. *Animals of the Arctic.* Holt, Rinehart & Winston, 1971.

MAMMALS
MECH, L. DAVID. *The Wolf.* Natural History Press, 1970.
PALMER, RALPH S. *The Mammal Guide.* Doubleday, 1954.
PERRY, RICHARD. *The World of the Polar Bear.* University of Washington Press, 1967.
PERRY, RICHARD. *The World of the Walrus.* Taplinger, 1968.
PETERSON, RANDOLPH L. *The Mammals of Eastern Canada.* Oxford

Press, Toronto, 1966.
RUE, LEONARD LEE III. *The World of the Beaver.* Lippincott, 1964.
SCHEFFER, VICTOR B. *Seals, Sea Lions and Walruses.* Stanford University Press, 1958.

BIRDS
GABRIELSON, IRA N., and F. C. LINCOLN. *The Birds of Alaska.* Stackpole, 1959.
GODFREY, W. EARL. *The Birds of Canada.* Queen's Printer, Ottawa, 1966.
LANSDOWNE, JAMES F., with JOHN A. LIVINGSTON. *Birds of the Northern Forest.* Houghton Mifflin, 1966.
MC NULTY, FAITH. *The Whooping Crane.* Dutton, 1966.
PETERS, HAROLD S., and THOMAS D. BURLEIGH. *The Birds of Newfoundland.* Houghton Mifflin, 1951.
SNYDER, L. L. *Arctic Birds of Canada.* University of Toronto Press, 1957.
TODD, W. E. CLYDE. *Birds of the Labrador Peninsula and Adjacent Areas.* University of Toronto Press, 1963.

PLANTS
BRITTON, MAX E. *Vegetation of the Arctic Tundra.* Oregon State University Press, 1966.
HALE, MASON E. *Lichen Handbook.* Smithsonian Institution, 1961.
POLUNIN, N. *Circumpolar Arctic Flora.* Oxford University Press, 1959.
PORSILD, A. E. *Illustrated Flora of the Canadian Arctic Archipelago.* Bulletin 146, National Museum of Canada, 1957.
PORSILD, A. E. *Plant Life in the Arctic.* National Museum of Canada. Misc. Publ. 74, 1951.
ROWE, J. S. *Forest Regions of Canada.* Bulletin 123, Canada, Department of Northern Affairs and Natural Resources, 1959.

WIGGINS, IRA L., and JOHN HUNTER THOMAS. *A Flora of the Alaskan Arctic Slope.* University of Toronto Press, 1962.

GEOGRAPHY AND GEOLOGY
BIRD, JOHN BRIAN. *The Physiography of Arctic Canada.* Johns Hopkins Press, 1967.
BROWN, ROGER J. E. *Permafrost in Canada.* University of Toronto Press, 1970.
FLINT, RICHARD F. *Glacial and Pleistocene Geology.* Wiley, 1957.
KIMBLE, GEORGE H. T., and DOROTHY GOOD. *Geography of the Northlands.* Wiley, 1955.
SUSLOV, SERGEI P. *Physical Geography of Asiatic Russia.* Freeman, 1961.
WILLIAMS, HOWELL. *Landscapes of Alaska, Their Geologic Evolution.* University of California Press, 1958.

CONSERVATION
BLADEN, VINCENT W. (Editor). *Canadian Population and Northern Colonization.* University of Toronto Press, 1962.
DAWSON, CARL A. (Editor). *The New North-West.* University of Toronto Press, 1947.
DUNBAR, MAXWELL J. *Ecological Development in Polar Regions.* Prentice-Hall, 1968.
LAYCOCK, GEORGE. *Alaska, the Embattled Frontier.* Houghton Mifflin, 1971.
MACDONALD, RONALD ST. JOHN (Editor). *The Arctic Frontier.* University of Toronto Press, 1966.
ROGERS, GEORGE W. (Editor). *Change in Alaska: People, Petroleum and Politics.* University of Washington Press, 1970.
UNDERHILL, FRANK H. (Editor). *The Canadian Northwest: Its Potentialities.* University of Toronto Press, 1959.

Illustration Credits and Acknowledgments

COVER: Moose and Mt. McKinley, Willis Peterson

ENDPAPERS: Steve McCutcheon

UNCAPTIONED PHOTOGRAPHS: 8–9: Tundra in summer, Willis Peterson 96–97: River in snow, John S. Crawford 154–155: Alpine arnica, Suzy Loder

ALL OTHER ILLUSTRATIONS: 10: George Herben, Van Cleve Photography 11: Annan Photo Features 12: Jen and Des Bartlett, Bruce Coleman, Inc. 13: Erwin A. Bauer 14–15: Douglas H. Pimlott 16: Alvin E. Staffan 17: C. G. Hampson 18: Charlie Ott, National Audubon Society 19: Matthew Kalmenoff 21: Richard H. Russell 22: Steve McCutcheon 22–23: Ernie Kuyt 24: F. W. Taylor 25: Ernie Kuyt 26–27: Steven C. Wilson 28–29: Ernie Kuyt 30–31: Fred Bruemmer 32: Tom Willock 33: John P. Milton 34–35: Fred Bruemmer 36: Annan Photo Features 37: Bob and Ira Spring 38–39: Josef Muench 40–41: Annan Photo Features 42–43: Michael Wotton 44: Fred Bruemmer 45: Alvin E. Staffan 46: Ernie Kuyt 48–49: John P. Milton 50: R. D. Muir 51: Richard Wright 52–53: Charlie Ott 54: Ernie Kuyt 55: Fred Bruemmer 56: F. W. Taylor 57–59: Steven C. Wilson 60–61: Averill S. Thayer, U.S. Department of the Interior 61: Fred Bruemmer 62: R. D. Muir 63–64: Philip S. Taylor 66–67: T. Pearce, Bruce Coleman, Inc. 68–69: Pierre Lamothe 70: Fred Bruemmer 71: Sven Samelius 72: Charles Fracé 73: D. Zingel, Bruce Coleman, Inc. 74: Fred Bruemmer 75: C. G. Hampson 76–77: Richard Kerbes 78–79: Joe Phillips 81: Fred Bruemmer 82–83: T. Pearce, Bruce Coleman, Inc. 84–85: Mary M. Tremaine 86–87: Karl W. Kenyon, National Audubon Society 88: Fred Bruemmer 89: Erwin A. Bauer 90: Patricia C. Henrichs 91: Fred Baldwin, National Audubon Society 92: Lee Miller, Photo Researchers, Inc. 93: Fred Bruemmer 94: Donald Cornelius 98–99: Patricia C. Henrichs 100: Erwin A. Bauer 101: Matthew Kalmenoff 102: Charlie Ott 102–103: Ted Loder 104: C. G. Hampson 105: Charlie Ott, National Audubon Society 106–107: Jen and Des Bartlett, Bruce Coleman, Inc. 107: J. Simon, Van Cleve Photography 108–109: Jen and Des Bartlett, Bruce Coleman, Inc. 110: Charles Fracé 111: Alvin E. Staffan 112–113: Dave Mech 114–115: Douglas H. Pimlott 115: Philip S. Taylor 116: Dave Mech; Fred Bruemmer 117: Annan Photo Features; Dave Mech; C. G. Hampson 118–119: Matthew Kalmenoff 120–121: John S. Crawford 122: Fred Bruemmer 123: Matthew Kalmenoff 124–125: Fred Bruemmer 126: Charles Fracé 127: F. W. Taylor 128: Patricia C. Henrichs 129–131: Charlie Ott, National Audubon Society 132–133: Bob and Ira Spring 134: Matthew Kalmenoff 135: Pierre Lamothe 136–137: Charles Fracé 138: G. Ronn, Ostman Agency 139: Harry Engels 140: Charlie Ott, National Audubon Society 141: John S. Crawford 142: T. Charles Dauphine, Jr. 143: Ben Strickland, Van Cleve Photography 144: R. D. Muir 145: F. W. Taylor 146: Steve McCutcheon 147: Ted Loder 148: Annan Photo Features 149: Steven C. Wilson 150–151: T. Charles Dauphine, Jr. 151: Steven C. Wilson 152: Grant Heilman 156–157: Pierre Lamothe 158: Ted Loder 159: Charlie Ott, National Audubon Society 160: Steve McCutcheon 161: Charlie Ott 162–163: Steve McCutcheon 164: Pierre Lamothe; Philip S. Taylor 165: Philip S. Taylor; Eric Hosking, Bruce Coleman, Inc. 166–167: Richard Kerbes 168: Ian A. McLaren 169: Fred Bruemmer 170: Michael Wotton 171: Annan Photo Features 172–173: Steve McCutcheon 174: Kenneth Fink, National Audubon Society; Willis Peterson 175: George Herben; Leonard Lee Rue III 176: Willis Peterson 177: Fred Bruemmer 178: John P. Milton 179–181: C. G. Hampson 182: Richard Kerbes 182–183: Karl Maslowski 184: Ian A. McLaren 185: Michael Wotton 186–187: Bjorn Reese, Smithsonian, Smithsonian National Associates, 1971 188–189: David O. Hill 190: Pierre Lamothe 191: Michael Wotton 192: Fred Bruemmer 193: Matthew Kalmenoff 194: C. David Repp, National Audubon Society 195: T. Charles Dauphine, Jr. 197: Willis Peterson 198: Fred Bruemmer 201: Patricia C. Henrichs 203–214: Charles Fracé 215–219: F. W. Taylor 220–221: Charles Fracé

PHOTO EDITOR: BARBARA KNOWLTON

ACKNOWLEDGMENTS: *The publisher and authors wish to thank Dr. J. P. Kelsall of the Canadian Wildlife Service and Dr. R. B. Weeden of the University of Alaska, both of whom through their special knowledge offered valuable assistance in the preparation of this book. Additional appreciation is due C. Gordon Fredine and Dr. Theodore W. Sudia of the National Park Service for reading the manuscript and contributing many useful suggestions.*

Index

[Page numbers in **boldface** type indicate reference to illustrations.]

232

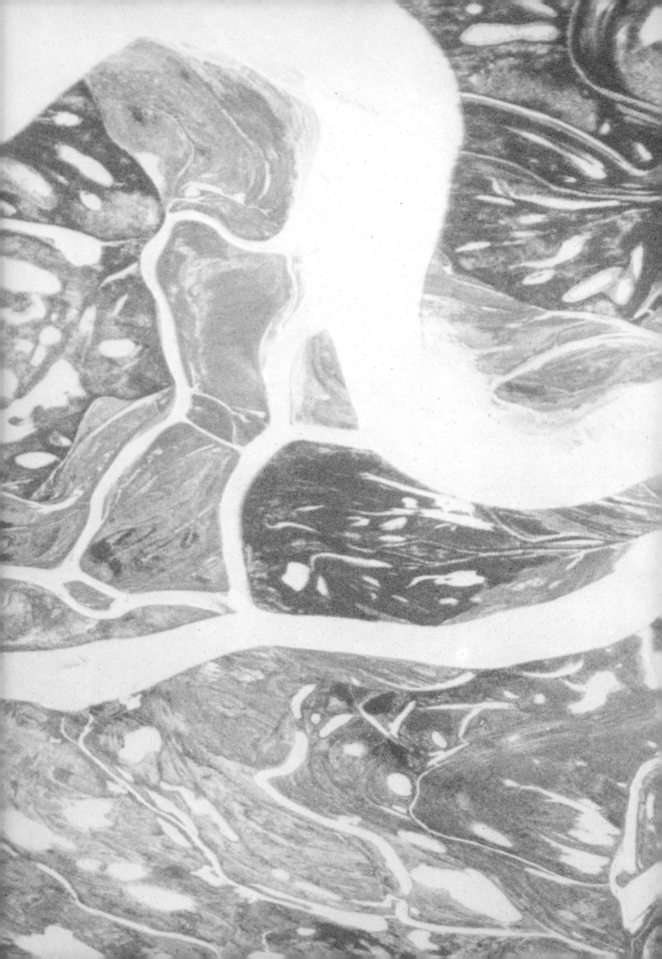